ASE Test Preparation

Automobile Certification Series

Electrical/Electronic Systems (A6)

5th Edition

DELMAR
CENGAGE Learning

Australia • Brazil • Japan • Korea • Mexico • Singapore • Spain • United Kingdom • United States

ASE Test Preparation: Automobile Certification Series, Electrical/Electronic Systems (A6), 5th Edition

Vice President, Technology and Trades Professional Business Unit: Gregory L. Clayton

Director, Professional Transportation Industry Training Solutions: Kristen L. Davis

Product Manager: Lori Bonesteel

Editorial Assistant: Danielle Filippone

Director of Marketing: Beth A. Lutz

Marketing Manager: Jennifer Barbic

Senior Production Director: Wendy Troeger

Senior Art Director: Benjamin Gleeksman

Content Project Management: PreMediaGlobal

Section Opener Image: Image Copyright Creations, 2012. Used under License from Shutterstock.com

ISBN-13: 978-1-111-12708-4

ISBN-10: 1-111-12708-5

Delmar Cengage Learning
5 Maxwell Drive
Clifton Park, NY 12065-2919
USA

Cengage Learning is a leading provider of customized learning solutions with office locations around the globe, including Singapore, the United Kingdom, Australia, Mexico, Brazil, and Japan. Locate your local office at: **international.cengage.com/region**

Cengage Learning products are represented in Canada by Nelson Education, Ltd.

For more information on transportation titles available from Delmar Cengage Learning, please visit our website at **www.trainingbay.cengage.com**

For more learning solutions, please visit our corporate website at **www.cengage.com**

Notice to the Reader

Publisher does not warrant or guarantee any of the products described herein or perform any independent analysis in connection with any of the product information contained herein. Publisher does not assume, and expressly disclaims, any obligation to obtain and include information other than that provided to it by the manufacturer. The reader is expressly warned to consider and adopt all safety precautions that might be indicated by the activities described herein and to avoid all potential hazards. By following the instructions contained herein, the reader willingly assumes all risks in connection with such instructions. The publisher makes no representations or warranties of any kind, including but not limited to, the warranties of fitness for particular purpose or merchantability, nor are any such representations implied with respect to the material set forth herein, and the publisher takes no responsibility with respect to such material. The publisher shall not be liable for any special, consequential, or exemplary damages resulting, in whole or part, from the readers' use of, or reliance upon, this material.

Printed in the United States of America
5 6 7 15

Table of Contents

Preface

Delmar, a part of Cengage Learning, is very pleased that you have chosen to use our ASE Test Preparation Guide to help prepare yourself for the Electrical/Electronic Systems (A6) ASE Certification Examination. This guide is designed to help prepare you for your actual exam, by providing you with an overview and introduction of the testing process, introducing you to the task list for the Electrical/Electronic Systems (A6) certification exam, giving you an understanding of what knowledge and skills you are expected to have in order to successfully perform the duties associated with each task area, and providing you with several preparation exams designed to emulate the live exam content in hopes of assessing your overall exam readiness.

If you have a basic working knowledge of the discipline you are testing for, you will find this book will be an excellent guide, helping you to understand the "must know" items needed to successfully pass the exam. This book is not a textbook. Its objective is to prepare the technician who has the requisite experience and schooling to take on the challenge of the ASE certification process. This guide cannot replace the hands-on experience and theoretical knowledge required by ASE to master the vehicle repair technology associated with this exam. If you are unable to understand more than a few of the preparation questions and their corresponding explanations in this book, it could be that you require either more shop-floor experience or further study.

This book begins by providing an overview of, and introduction to, the testing process. This section outlines what we recommend you do to prepare, what to expect on the actual test day, and overall methodologies for your success. This section is followed by a detailed overview of the ASE task list to include explanations of the knowledge and skills you must possess to successfully answer questions related to each particular task. After the task list, we provide six sample preparation exams for you to use as a means of evaluating areas of understanding, as well as areas requiring improvement in order to successfully pass the ASE exam. Delmar is the first and only test preparation organization to provide so many unique preparation exams. We enhanced our guides to include this support as a means of providing you with the best preparation product available. Section 6 of this guide includes the answer keys for each preparation exam, along with the answer explanations for each question. Each answer explanation also contains a reference back to the related task or tasks that it assesses. This will provide you with a quick and easy method for referring back to the task list whenever needed. The last section of this book contains blank answer sheet forms you can use as you attempt each preparation exam, along with a glossary of terms.

OUR COMMITMENT TO EXCELLENCE

Thank you for choosing Delmar, Cengage Learning for your ASE test preparation needs. All of the writers, editors, and Delmar staff have worked very hard to make this test preparation guide second to none. We feel confident that you will find this guide easy to use and extremely beneficial as you prepare for your actual ASE exam.

Delmar, Cengage Learning has sought out the best subject matter experts in the country to help with the development of this fifth edition of the *ASE Test Preparation: Automobile Certification Series, Electrical/Electronic Systems (A6), 5th Edition*. Preparation questions are authored and then reviewed by a group of certified, subject-matter experts to ensure the highest level of quality and validity to our product.

If you have any questions concerning this guide or any guide in this series, please visit us on the web at **http://www.trainingbay.cengage.com**.

For web-based online test preparation for ASE certifications, please visit us on the web at **http://www.techniciantestprep.com/** to learn more.

ABOUT THE AUTHOR

Jerry Clemons has been around cars, trucks, equipment, and machinery throughout his whole life. Being raised on a large farm in central Kentucky provided him with an opportunity to complete mechanical repair procedures from an early age. Jerry earned an Associate in Applied Science degree in Automotive Technology from Southern Illinois University and a Bachelor of Science degree in Vocational, Industrial, and Technical Education from Western Kentucky University. Jerry also has completed a Master of Science degree in Safety, Security, and Emergency Management from Eastern Kentucky University. Jerry has been employed at Elizabethtown Community and Technical College since 1999 and is currently an Associate Professor for the Automotive and Diesel Technology Programs. Jerry holds the following ASE certifications: Master Medium/Heavy Truck Technician, Master Automotive Technician, Advanced Engine Performance (L1), Truck Equipment Electrical Installation (E2), and Automotive Service Consultant (C1). Jerry is a member of the Mobile Air Conditioning Society (MACS) as well as a member of the North American Council of Automotive Teachers (NACAT). Jerry has been involved in developing transportation material for Cengage Learning for seven years.

ABOUT THE SERIES ADVISOR

Mike Swaim has been an Automotive Technology Instructor at North Idaho College, Coeur d'Alene, Idaho since 1978. He is an Automotive Service Excellence (ASE) Certified Master Technician since 1974 and holds a Lifetime Certification from Mobile Air Conditioning Society. He served as Series Advisor to all nine of the 2011 Automobile/Light Truck Certification Tests (A Series) of Cengage, Delmar ASE Test Preparation titles, and is the author of *ASE Test Preparation: Automobile Certification Series, Undercar Specialist Designation (X1), 5th Edition.*

SECTION 1

The History and Purpose of ASE

ASE began as the National Institute for Automotive Service Excellence (NIASE). It was founded as a non-profit, independent entity in 1972 by a group of industry leaders with the single goal of providing a means for consumers to distinguish between incompetent and competent technicians. It accomplishes this goal through the testing and certification of repair and service professionals. Though it is still known as the National Institute for Automotive Service Excellence, it is now called "ASE" for short.

Today, ASE offers more than 40 certification exams in automotive, medium/heavy duty trucks, collision repair and refinish, school bus, transit bus, parts specialist, automobile service consultant, and other industry-related areas. At this time, there are more than 385,000 professionals nationwide with current ASE certifications. These professionals are employed by new car and truck dealerships, independent repair facilities, fleets, service stations, franchised service facilities, and more.

ASE's certification exams are industry-driven and cover practically every on-highway vehicle service segment. The exams are designed to stress the knowledge of job-related skills. Certification consists of passing at least one exam and documenting two years of relevant work experience. To maintain certification, those with ASE credentials must be re-tested every five years.

While ASE certifications are a targeted means of acknowledging the skills and abilities of an individual technician, ASE also has a program designed to provide recognition for highly qualified repair, support, and parts businesses. The Blue Seal of Excellence Recognition Program allows businesses to showcase their technicians and their commitment to excellence. One of the requirements of becoming Blue Seal recognized is that the facility must have a minimum of 75 percent of their technicians ASE certified. Additional criteria apply, and program details can be found on the ASE website.

ASE recognized that educational programs serving the service and repair industry also needed a way to be recognized as having the faculty, facilities, and equipment to provide a quality education to students wanting to become service professionals. Through the combined efforts of ASE, industry, and education leaders, the non-profit organization entitled the National Automotive Technicians Education Foundation (NATEF) was created in 1983 to evaluate and recognize academic programs. Today more than 2,000 educational programs are NATEF certified.

For additional information about ASE, NATEF, or any of their programs, the following contact information can be used:

National Institute for Automotive Service Excellence (ASE)
101 Blue Seal Drive S.E.
Suite 101
Leesburg, VA 20175
Telephone: 703-669-6600
Fax: 703-669-6123
Website: **www.ase.com**

SECTION 2

Overview and Introduction

Participating in the National Institute for Automotive Service Excellence (ASE) voluntary certification program provides you with the opportunity to demonstrate that you are a qualified and skilled professional technician who has the "know-how" required to successfully work on today's modern vehicles.

EXAM ADMINISTRATION

Note: After November 2011, ASE will no longer offer paper and pencil certification exams. There will be no Winter testing window in 2012, and ASE will offer and support CBT testing exclusively starting in April 2012.

ASE provides computer-based testing (CBT) exams, which are administered at test centers across the nation. It is recommended that you go to the ASE website at ***http://www.ase.com*** and review the conditions and requirements for this type of exam. There is also an exam demonstration page that allows you to personally experience how this type of exam operates before you register.

CBT exams are available four times annually, for two-month windows, with a month of no testing in between each testing window:

- January/February – Winter testing window
- April/May – Spring testing window
- July/August – Summer testing window
- October/November – Fall testing window

Please note, testing windows and timing may change. It is recommended you go to the ASE web site at ***http://www.ase.com*** and review the latest testing schedules.

UNDERSTANDING TEST QUESTION BASICS

ASE exam questions are written by service industry experts. Each question on an exam is created during an ASE-hosted "item-writing" workshop. During these workshops, expert service representatives from manufacturers (domestic and import), aftermarket parts and equipment manufacturers, working technicians, and technical educators gather to share ideas and convert them into actual exam questions. Each exam question written by these experts then must survive review by all members of the group. The questions are designed to address the practical application of repair and diagnosis knowledge and skills practiced by technicians in their day-to-day work.

After the item-writing workshop, all questions are pretested and quality-checked on a national sample of technicians. Those questions that meet ASE standards of quality and accuracy are included in the scored sections of the exams; the "rejects" are sent back to the drawing board or discarded altogether.

Depending on the topic of the certification exam, you will be asked between 40 and 80 multiple-choice questions. You can determine the approximated number of questions you can expect to be asked during the Electrical/Electronic Systems (A6) certification exam by reviewing the task list in Section 4 of this book. The five-year recertification exam will cover this same content; however, the number of questions for each content area of the recertification exam will be reduced by approximately one-half.

> *Note:* Exams may contain questions that are included for statistical research purposes only. Your answers to these questions will not affect your score, but since you do not know which ones they are, you should answer all questions in the exam.

Using multiple criteria, including cross-sections by age, race, and other background information, ASE is able to guarantee that exam questions do not include bias for or against any particular group. A question that shows bias toward any particular group is discarded.

TEST-TAKING STRATEGIES

Before beginning your exam, quickly look over the exam to determine the total number of questions that you will need to answer. Having this knowledge will help you manage your time throughout the exam to ensure you have enough available to answer all of the questions presented. Read through each question completely before marking your answer. Answer the questions in the order they appear on the exam. Leave the questions blank that you are not sure of and move on to the next question. You can return to those unanswered questions after you have finished the others. These questions may actually be easier to answer at a later time after your mind has had additional time to consider them on a subconscious level. In addition, you might find information in other questions that will help you recall the answers to some of them.

Multiple-choice exams sometimes are challenging because often several choices may seem possible, or partially correct, and therefore, it may be difficult to decide on the most appropriate answer choice. The best strategy, in this case, is to first determine the correct answer before looking at the answer options. If you see the answer you decided on, you should still be careful to examine the other answer options to make sure that none seems more correct than yours. If you do not know or are not sure of the answer, read each option very carefully and try to eliminate those options that you know are incorrect. That way, you often can arrive at the correct choice through a process of elimination.

If you have gone through the entire exam, and you still do not know the answer to some of the questions, *then guess*. Yes, guess. You then have at least a 25 percent chance of being correct. Although your score is based on the number of questions answered correctly, any question left blank, or unanswered, is automatically scored as incorrect.

There is a lot of "folk" wisdom on the subject of test taking that you may hear about as you prepare for your ASE exam. For example, some people would advise you to avoid response options that use certain words such as *all, none, always, never, must,* and *only,* to name a few. This, they claim, is because nothing in life is exclusive. They would advise you to choose response options that use words that allow for some exception, such as *sometimes, frequently, rarely, often, usually, seldom,* and *normally.* They would also advise you to avoid the first and last option (A or D) because exam writers, they feel, are more comfortable if they put the correct answer in the middle (B or C) of the choices. Another recommendation often offered is to select the option that is either shorter or longer than the other three choices because it is more likely to be correct. Some would advise you to never change an answer since your first intuition is usually correct. Another area of "folk" wisdom focuses specifically on any repetitive patterns created by your question responses (e.g., A, B, C, A, B, C, A, B, C).

Many individuals may say that actual grains of truth are in this "folk" wisdom, and whereas with some exams, this may prove true, it is not relevant in regard to the ASE certification exams. ASE validates all exam questions and test forms through a national sample of technicians, and only those questions and test forms that meet ASE standards of quality and accuracy are included in the scored sections of the exams. Any biased questions or patterns are discarded altogether, and therefore, it is highly unlikely you will experience any of this "folk" wisdom on an actual ASE exam.

PREPARING FOR THE EXAM

Delmar, Cengage Learning wants to make sure we are providing you with the most thorough preparation guide possible. To demonstrate this, we have included hundreds of preparation questions in this guide. These questions are designed to provide as many opportunities as possible to prepare you to successfully pass your ASE exam. The preparation approach we recommend and outline in this book is designed to help you build confidence in demonstrating what task area content you already know well while also outlining what areas you should review in more detail prior to the actual exam.

We recommend that your first step in the preparation process should be to thoroughly review Section 3 of this book. This section contains a description and explanation of the type of questions you will find on an ASE exam.

After you understand how the questions will be presented, we then recommend that you thoroughly review Section 4 of this book. This section contains information that will help you establish an understanding of what the exam will be evaluating, and specifically, how many questions to expect in each specific task area.

As your third preparatory step, we recommend you complete your first preparation exam, located in Section 5 of this book. Answer one question at a time. After you answer each question, review the answer and question explanation information located in Section 6. This section will provide you with instant feedback, allowing you to gauge your progress, one question at a time, throughout this first preparation exam. If after reading the question explanation you do not feel you understand the reasoning for the correct answer, go back and review the task list overview (Section 4) for the task that is related to that question. Included with each question explanation is a clear identifier of the task area that is being assessed (e.g., Task A.1). If at that point you still do not feel you have a solid understanding of the material, identify a good source of information on the topic, such as an educational course, textbook, or other related source of topical learning, and do some additional studying.

After you have completed your first preparation exam and have reviewed your answers, you are ready to complete your next preparation exam. A total of six practice exams are available in Section 5 of this book. For your second preparation exam, we recommend that you answer the questions as if you were taking the actual exam. Do not use any reference material or allow any interruptions in order to get a feel for how you will do on the actual exam. Once you have answered all of the questions, grade your results using the Answer Key in Section 6. For every question that you gave an incorrect answer to, study the explanations to the answers and/or the overview of the related task areas. Try to determine the root cause for missing the question. The easiest thing to correct is learning the correct technical content. The hardest things to correct are behaviors that lead you to an incorrect conclusion. If you knew the information but still answered the question incorrectly, there is likely a test-taking behavior that will need to be corrected. An example of this would be reading too quickly and skipping over words that affect your reasoning. If you can identify what you did that caused you to answer the question incorrectly, you can eliminate that cause and improve your score.

Here are some basic guidelines to follow while preparing for the exam:

- Focus your studies on those areas in which you are weak.
- Be honest with yourself when determining whether you understand something.
- Study often but for short periods of time.
- Remove yourself from all distractions when studying.
- Keep in mind that the goal of studying is not just to pass the exam; the real goal is to learn.
- Prepare physically by getting a good night's rest before the exam and eat meals that provide energy but do not cause discomfort.
- Arrive early to the exam site to avoid long waits as test candidates check in.
- Use all of the time available for your exams. If you finish early, spend the remaining time reviewing your answers.
- Do not leave any questions unanswered. If absolutely necessary, guess. All unanswered questions are automatically scored as incorrect.

Here are some items you will need to bring with you to the exam site:

- A valid government or school-issued photo ID
- Your test center admissions ticket
- A watch (not all test sites have clocks)

> ***Note:*** Books, calculators, and other reference materials are not allowed in the exam room. The exceptions to this list are English-Foreign dictionaries or glossaries. All items will be inspected before and after testing.

WHAT TO EXPECT DURING THE EXAM

When taking a CBT exam, as soon as you are seated in the testing center, you will be given a brief tutorial to acquaint you with the computer-delivered test prior to taking your certification exam(s). The CBT exams allow you to select only one answer per question. You can also change your answers as many times as you like. When you select a second answer choice, the CBT will automatically unselect your first answer choice. If you want to skip a question to return to later, you can utilize the "flag" feature, which will allow you to quickly identify and review questions whenever you are ready. Prior to completing your exam, you will also be provided with an opportunity to review your answers and address any unanswered questions.

TESTING TIME

Each individual ASE CBT exam has a fixed time limit. Individual exam times will vary based upon exam area and will range anywhere from a half hour to two hours. You will also be given an additional 30 minutes beyond what is allotted to complete your exams to ensure you have adequate time to perform all necessary check-in procedures, complete a brief CBT tutorial, and potentially complete a post-test survey.

You can register for and take multiple CBT exams during one testing appointment. The maximum time allotment for a CBT appointment is four and a half hours. If you happen to register for so many exams that you will require more time than this, your exams will be scheduled into multiple appointments. This could mean that you have testing on both the morning and afternoon of the same day, or they could be scheduled on different days, depending on your personal preference and the test center's schedule.

It is important to understand that if you arrive late for your CBT test appointment, you will not be able to make up any missed time. You will only have the scheduled amount of time remaining in your appointment to complete your exam(s).

Also, while most people finish their CBT exams within the time allowed, others might feel rushed or not be able to finish the test, due to the implied stress of a specific, individual time limit allotment. Before you register for the CBT exams, you should review the number of exam questions that will be asked along with the amount of time allotted for that exam to determine whether you feel comfortable with the designated time limitation or not.

As an overall time management recommendation, you should monitor your progress and set a time limit you will follow with regard to how much time you will spend on each individual exam question. This should be based on the total number of questions you will be answering.

Also, it is very important to note that if for any reason you wish to leave the testing room during an exam, you must first ask permission. If you happen to finish your exam(s) early and wish to leave the testing site before your designated session appointment is completed, you are permitted to do so only during specified dismissal periods.

UNDERSTANDING HOW YOUR EXAM IS SCORED

You can gain a better perspective about the ASE certification exams if you understand how they are scored. ASE exams are scored by an independent organization having no vested interest in ASE or in the automotive industry.

Each question carries the same weight as any other question. For example, if there are 50 questions, each is worth 2 percent of the total score.

Your exam results can tell you:

- Where your knowledge equals or exceeds that needed for competent performance, or
- Where you might need more preparation.

Your ASE exam score report is divided into content "task" areas; it will show the number of questions in each content area and how many of your answers were correct. These numbers provide information about your performance in each area of the exam. However, because there may be a different number of questions in each content area of the exam, a high percentage of correct answers in an area with few questions may not offset a low percentage in an area with many questions.

It should be noted that one does not "fail" an ASE exam. The technician who does not pass is simply told "More Preparation Needed." Though large differences in percentages may indicate problem areas, it is important to consider how many questions were asked in each area. Since each exam evaluates all phases of the work involved in a service specialty, you should be prepared in each area. A low score in one area could keep you from passing an entire exam.

There is no such thing as average. You cannot determine your overall exam score by adding the percentages given for each task area and dividing by the number of areas. It does not work that way because there generally is not the same number of questions in each task area. A task area with 20 questions, for example, counts more toward your total score than a task area with 10 questions.

Your exam report should give you a good picture of your results and a better understanding of your strengths and areas needing improvement for each task area.

If you fail to pass the exam, you may take it again at any time it is scheduled to be administered. You are the only one who will receive your exam score. Exam scores will not be given over the telephone by ASE nor will they be released to anyone without your written permission.

SECTION 3

Types of Questions on an ASE Exam

Understanding not only what content areas will be assessed during your exam, but how you can expect exam questions to be presented will enable you to gain the confidence you need to successfully pass an ASE certification exam. The following examples will help you recognize the types of question styles used in ASE exams and assist you in avoiding common errors when answering them.

Most initial certification tests are made up of 40 to 80 multiple-choice questions. The five-year recertification exams will cover the same content as the initial exam; however, the actual number of questions for each content area will be reduced by approximately one-half. Refer to Section 4 of this book for specific details regarding the number of questions to expect during the initial Electrical/Electronic Systems (A6) certification exam.

Multiple-choice questions are an efficient way to test knowledge. To correctly answer them, you must consider each answer choice as a possibility, and then choose the answer choice that *best* addresses the question. To do this, read each word of the question carefully. Do not assume you know what the question is asking until you have finished reading the entire question.

About 10 percent of the questions on an actual ASE exam will reference an illustration. These drawings contain the information needed to correctly answer the question. The illustration should be studied carefully before attempting to answer the question. When the illustration is showing a system in detail, look over the system and try to figure out how the system works before you look at the question and the possible answers. This approach will ensure that you do not answer the question based upon false assumptions or partial data, but instead have reviewed the entire scenario being presented.

MULTIPLE-CHOICE/DIRECT QUESTIONS

The most common type of question used on an ASE exam is the direct multiple-choice style question. This type of question contains an introductory statement, called a stem, followed by four options: three incorrect answers, called distracters, and one correct answer, the key.

When the questions are written, the point is to make the distracters plausible to draw an inexperienced technician to inadvertently select one of them. This type of question gives a clear indication of the technician's knowledge.

Here is an example of a direct style question:

1. Which tool is recommended by manufacturers to perform voltage measurements on circuits that are controlled or monitored by a control module?

TASK A.1

   A. DMM
   B. Test
   C. Continuity tester
   D. Oscilloscope

Answer A is correct. A digital multi-meter (DMM) that has at least 10 megohms of resistance is the recommended tool to use on circuits that involve a control module. These tools will not cause any problems with electronic circuits because minimal electrical flow moves through the leads due to the high impedance in the meter.

Answer B is incorrect. A test light has very little resistance and could cause damage to a control module as well as cause a negative effect on the circuit being tested.

Answer C is incorrect. A continuity tester is never used to check voltage in a circuit. It is only used on an un-powered circuit to check for a positive electrical connection (continuity).

Answer D is incorrect. An oscilloscope is not required to perform a voltage measurement on a circuit. An oscilloscope is a useful tool that will show the voltage represented by a line on a screen (voltage over time).

COMPLETION QUESTIONS

A completion question is similar to the direct question except the statement may be completed by any one of the four options to form a complete sentence. Here is an example of a completion question:

TASK A.9

1. A technician will most likely refer to a wiring diagram to find:
   - A. The power and ground distribution for the circuit.
   - B. The location of the ground connection.
   - C. Updated factory information about pattern failures.
   - D. A flowchart for troubleshooting an electrical problem.

Answer A is correct. A wiring diagram will provide details about how the power and ground are connected to the circuit.

Answer B is incorrect. A wiring diagram does not typically provide the location of electrical components such as a ground connection.

Answer C is incorrect. A wiring diagram does not typically provide updated factory information. Technical service bulletins provide updated factory information about pattern failures.

Answer D is incorrect. A wiring diagram does not provide any flowcharts for troubleshooting.

TECHNICIAN A, TECHNICIAN B QUESTIONS

This type of question is usually associated with an ASE exam. It is, in fact, two true-false statements grouped together, such as: "Technician A says…" and "Technician B says…", followed by "Who is correct?"

In this type of question, you must determine whether either, both, or neither of the statements are correct. To answer this type of question correctly, you must carefully read each technician's statement and judge it on its own merit.

Sometimes this type of question begins with a statement about some analysis or repair procedure. This statement provides the setup or background information required to understand the conditions about which Technician A and Technician B are talking, followed by two statements about the cause of the concern, proper inspection, identification, or repair choices. Analyzing this type of question is a little easier than the other types because there are only two ideas to consider, although there are still four choices for an answer.

Again, Technician A, Technician B questions are really double true-or-false questions. The best way to analyze this type of question is to consider each technician's statement separately. Ask yourself, "Is A true or false? Is B true or false?" Once you have completed an individual evaluation of each statement, you will have successfully determined the correct answer choice for the question, "Who is correct?"

An important point to remember is that an ASE Technician A, Technician B question will never have Technician A and B directly disagreeing with each other. That is why you must evaluate each statement independently.

An example of a Technician A/Technician B style question looks like this:

1. Technician A says that a 12 volt battery that has 6 volts at the posts is 50 percent charged. Technician B says that a 12 volt battery that has 12.6 volts at the posts is overcharged. Who is correct?

TASK B.1

   A. A only
   B. B only
   C. Both A and B
   D. Neither A nor B

Answer A is incorrect. The 50 percent charge level on a 12 volt battery is 12.2 volts.

Answer B is incorrect. A battery with 12.6 volts at the posts is fully charged.

Answer C is incorrect. Neither Technician is correct.

Answer D is correct. Neither Technician is correct. A battery that has only 6 volts at the posts is severely discharged. A fully charged automotive battery should have 12.6 volts. These batteries have 6 cells that produce 2.1 volts each.

EXCEPT QUESTIONS

Another type of question used on ASE exams contains answer choices that are all correct except for one. To help easily identify this type of question, whenever it is presented in an exam, the word "EXCEPT" will always be displayed in capital letters. Furthermore, a cautionary statement will alert you to the fact that the next question is different from the ones otherwise found in the exam. With the EXCEPT type of question, only one incorrect choice will actually be listed among the options, and that incorrect choice will be the key to the question. That is, the incorrect statement is counted as the correct answer for that question.

Be careful to read these question types slowly and thoroughly; otherwise, you may overlook what the question is actually asking and answer the question by selecting the first correct statement.

An example of this type of question would appear as follows:

1. All of the following procedures are acceptable methods of locating excessive parasitic draw EXCEPT:

TASK A.7

   A. Disconnect the battery ground cable from the engine block.
   B. Remove the fuses one at a time while watching the ammeter.
   C. Inspect the whole vehicle for any lamp that could be staying on.
   D. Disconnect the charge wire at the generator while watching the ammeter.

Answer A is correct. Disconnecting the negative battery cable at the engine block will not assist in finding an excessive parasitic draw. All current flow will stop when the battery cable is removed.

Answer B is incorrect. Removing the fuses one at a time is an organized way to find an excessive parasitic draw.

Answer C is incorrect. Inspecting the vehicle for lamps that are staying on is a good step in diagnosing an excessive parasitic draw.

Answer D is incorrect. Disconnecting the charge wire at the generator is a step in diagnosing an excessive parasitic draw problem.

LEAST LIKELY QUESTIONS

LEAST LIKELY questions are similar to EXCEPT questions. Look for the answer choice that would be the LEAST LIKELY cause of the described situation. To help easily identify this type of question, whenever they are presented in an exam, the words "LEAST LIKELY" will always be displayed in capital letters. In addition, you will be alerted before a LEAST LIKELY question is posed. Read the entire question carefully before choosing your answer.

An example of this type of question is shown below:

TASK B.9

1. Which of the following functions is LEAST LIKELY to be performed by the starter solenoid?
   A. Prevents the armature from over-spinning
   B. Push the drive gear out to the flywheel
   C. Connects the "bat" terminal to the "motor" terminal
   D. Provides a path for high current to flow

Answer A is correct. The starter solenoid has no control over the speed of the armature.

Answer B is incorrect. The starter solenoid creates linear movement to push the drive gear into the flywheel.

Answer C is incorrect. The starter solenoid acts as a switch to connect the "bat" terminal to the "motor" terminal when the starter is engaged.

Answer D is incorrect. The starter solenoid provides an electrical path for high current flow through the contacts and into the starter housing.

SUMMARY

The question styles outlined in this section are the only ones you will encounter on any ASE certification exam. ASE does not use any other types of question styles, such as fill-in-the-blank, true/false, word-matching, or essay. ASE also will not require you to draw diagrams or sketches to support any of your answer selections, although any of the above described question styles may include illustrations, charts, or schematics to clarify a question. If a formula or chart is required to answer a question, it will be provided for you.

Task List Overview

SECTION 4

INTRODUCTION

This section of the book outlines the content areas or *task list* for this specific certification exam, along with a written overview of the content covered in the exam.

The task list describes the actual knowledge and skills necessary for a technician to successfully perform the work associated with each skill area. This task list is the fundamental guideline you should use to understand what areas you can expect to be tested on, as well as how each individual area is weighted to include the approximate number of questions you can expect to be given for that area during the ASE certification exam. It is important to note that the number of exam questions for a particular area is to be used as a guideline only. ASE advises that the questions on the exam may not equal the number specifically listed on the task list. The task lists are specifically designed to tell you what ASE expects you to know how to do and to help prepare you to be tested.

Similar to the role this task list will play in regard to the actual ASE exam, Delmar, Cengage Learning has developed six preparation exams, located in Section 5 of this book, using this task list as a guide. It is important to note that although both ASE and Delmar, Cengage Learning use the same task list as a guideline for creating these test questions, none of the test questions you will see in this book will be found in the actual, live ASE exams. This is true for any test preparatory material you use. Real exam questions are *only* visible during the actual ASE exams.

Task List at a Glance

The Electrical/Electronic Systems (A6) 2012 task list focuses on six core areas, and you can expect to be asked a total of approximately 50 questions on your certification exam, broken out as outlined:

A. General Electrical/Electronic System Diagnosis (13 questions)
B. Battery and Starting System Diagnosis and Repair (9 questions)
C. Charging System Diagnosis and Repair (5 questions)
D. Lighting Systems Diagnosis and Repair (6 questions)
E. Instrument Cluster and Driver Information Systems Diagnosis and Repair (6 questions)
F. Body Electrical Systems Diagnosis and Repair (11 questions)

Based upon this information, the following graphic is a general guideline demonstrating which areas will have the most focus on the actual certification exam. This data may help you prioritize your time when preparing for the exam.

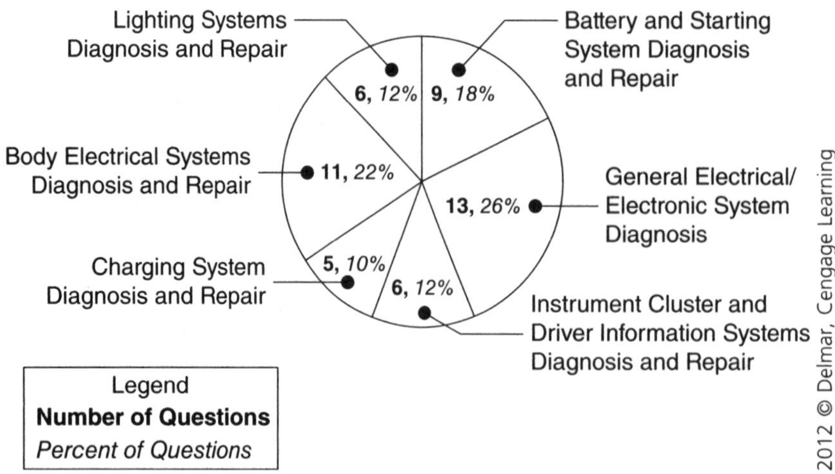

Note: There could be additional questions that are included for statistical research purposes only. Your answers to these questions will not affect your test score, but since you do not know which ones they are, you should answer all questions in the test. The five-year Recertification Test will cover the same content areas as those listed above. However, the number of questions in each content area of the Recertification Test will be reduced by one-half.

A. General Electrical/Electronic System Diagnosis (13 Questions)

1. Check voltages and voltage drops in electrical/ electronic circuits; interpret readings and determine needed repairs.

A voltmeter may be connected across a component in a circuit to measure the voltage drop across the component. Current must be flowing through the circuit during the voltage drop test. The amount of voltage drop depends on the resistance in the component and the amount of current flow the component requires. In a normal electrical circuit, the electrical load should drop most of the voltage; and the wires, connectors, switches, and relays should drop very little voltage. Performing voltage drop tests on the wires, connectors, switches and relays while the circuit is energized is a very good way to determine whether these items are operating correctly.

2. Check current flow in electrical/electronic circuits and components; interpret readings and determine needed repairs.

An ammeter has low internal resistance so that it can measure current through it. The meter must be connected in series with a circuit. Some ammeters have an inductive clamp that fits over a wire in the circuit. These ammeters measure the current flow from the strength of the magnetic field surrounding the wire. High current flow is caused by high voltage or low resistance. On the other hand, low current flow results from high resistance or low voltage. Since current is the supply to get any electrical job done, an ammeter comes in handy for many types of diagnosis. Here are some examples of things a technician would use an ammeter to measure.

Component	Why Use the Ammeter
Generator/ Alternator	The vehicle's charging system is the power station for everything electrical. Generators are matched to the total load the vehicle might generate so that it can maintain adequate voltage while supplying enough current. Technicians use an ammeter, usually in conjunction with a voltmeter, to measure the ability of the generator to meet its intended output specification.
Battery	The battery stores current as potential under a certain pressure (voltage). An ammeter is used, usually in conjunction with a voltmeter, to measure the potential of the battery when it is working to see whether it meets its intended load specification.
High-Power Circuits	Electric fuel pumps, HVAC blower motors, starters, electric cooling fans, window motors, solenoids, and any other high load item can be tested in the same way to determine whether they require more current than they were designed to. In contrast to the battery and the generator, these items consume the power generated by the charging system. In many cases, a technician can use an ammeter to not only find a component that has an electrical issue, but also to find out whether the system it operates has a mechanical issue. An example would be a window motor that blows fuses due to an excessive demand for current. It could be that the motor is good, but that the window is stuck and will not move freely causing high current demand on a circuit. A sharp technician must keep this in mind before replacing a component that is not faulty. Some internal engine problems are missed until after a starter that needs 600 amps to turn the engine over is replaced.

3. Check continuity and resistances in electrical/electronic circuits and components; interpret readings and determine needed repairs.

An ohmmeter has an internal power source. Meter damage may result if it is connected to a live circuit. Most currently produced ohmmeters are auto-ranging (automatically selecting the proper resistance range). If not, be sure to select the proper range when measuring a circuit. For example, when a component has resistance of 10,000 ohms, select the X1000 or 10K range.

Some circuits contain components that have published resistance specs. This is where an ohmmeter comes in handy. Fuel injectors, solenoids, and sensors are examples of items that often have specifications for resistance, which can be a good way to determine whether a component has a problem.

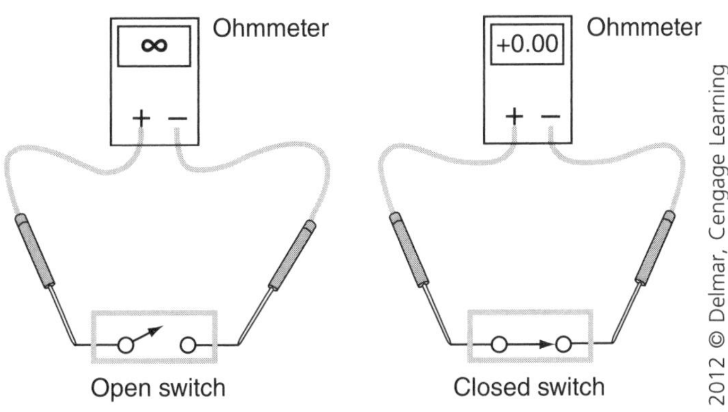

4. Check electronic circuit waveforms; interpret readings and determine needed repairs.

An oscilloscope converts electrical signals into a visual image representing voltage changes over a specific period of time. An upward movement of the trace means the voltage has increased, whereas a downward movement means the voltage has decreased. If the trace stays flat, the voltage is staying at that level. As the trace moves across the screen of the oscilloscope, time is represented.

The size and clarity of the trace is dependent on the voltage scale and the time reference selected by the technician. Most scopes are equipped with controls that allow voltage and time interval selection.

Analog scopes show the actual activity of a circuit and are referred to as "real-time" scopes. Digital scopes convert the voltage signals into digital signals; therefore, some delay between the electrical activity and the display is experienced. This delay does give the trace a cleaner appearance, as an analog trace is clean only when the voltage has been constant for some time.

The following is an example of a waveform captured from a digital oscilloscope.

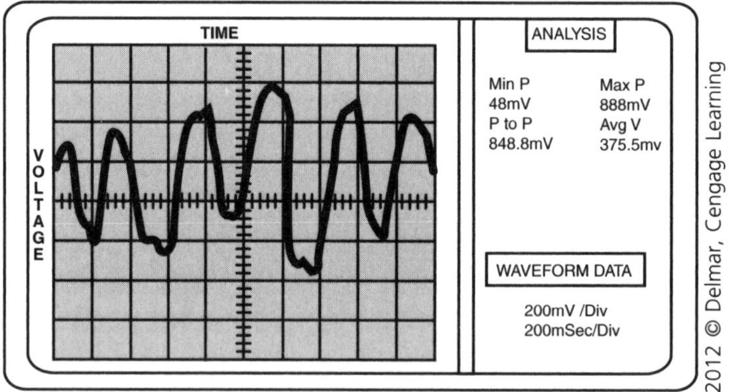

5. Use scan-tool data and/or diagnostic trouble codes (DTCs) to diagnose electronic systems; interpret readings and determine necessary action.

A scan tool is designed to communicate with the vehicle's computer. Connected to the computer through a diagnostic connector, a scan tool can access live data and trouble codes as well as run tests to check system operations. Many scan tools have the ability to set up special functions within the vehicles' computer systems that help a technician test circuits or to activate them so that the technician can watch the scan tool data for changes. This is known as bidirectional communication. Many of the electronic systems of late-model vehicles can be accessed and even exercised by the scan tool.

The data displayed on a scan tool should be compared to the expected or known good data tables given in the service manual for the vehicle being tested. Depending on the year of the vehicle and the system with which it is equipped, the trouble codes displayed on the scan tool may indicate a general problem area or may identify a specific part of a circuit. Observing the activity of various inputs and outputs can lead a technician to problems that may be related or unrelated to the computer system.

Diagnostic trouble codes (DTCs) can be retrieved with a scan tool. DTCs are a function of most of the control modules on late-model vehicles. Each control module is equipped with diagnostic software that will record a DTC when a problem is present for a short period of time. The technician can retrieve the DTC and then follow a diagnostic routine to find out what caused the DTC. After the repair is made, the DTC should be cleared from the memory of the control module with the scan tool.

6. Find shorts, grounds, opens, and resistance problems in electrical/electronic circuits; determine needed repairs.

Following is a list of the various circuit problems that can occur in electrical circuits.

- Short circuit: A short circuit is when the electrical path is changed. All circuits are designed to be routed through the various components such as wires, switches, fuses, relays, motors, lights, and heat strips. A short circuit happens when the electrical flow takes an unwanted path. A voltmeter is the preferred tool to use to locate short circuits.
- High-resistance circuit: Circuits that have high resistance as a result of corrosion, loose terminals, and burnt contacts will have reduced current flow, and the electrical components will not function correctly. These circuit problems can be found by using either an ohmmeter or a voltmeter. The circuit must be disconnected from electrical power before using an ohmmeter. The ohmmeter is connected in parallel with the part of the circuit that needs to be tested. A voltmeter can be used to check for circuit problems as well. The circuit must be energized when performing a voltage drop test. The voltmeter is connected in parallel with the circuit while the power is turned on.
- Grounded circuit: A grounded circuit is sometimes called a short to ground. If an electrical circuit gets a short to ground on the power side of the circuit, the current flow will rapidly increase and probably open the circuit protection device. If a short to ground happens in the ground side of the circuit, the circuit will not be changed. The exception to this example would be if the circuit is switched on the ground side. In this case, the circuit would operate all of the time. An ohmmeter can be used to locate grounded circuits. Remember that the power must be turned off when using an ohmmeter.
- Open circuit: An open circuit does not allow current to flow. An open circuit will not open any circuit protection device. An ohmmeter or a voltmeter can be used to locate open circuits. On an unpowered circuit, the ohmmeter can be connected in parallel with the part of the circuit that needs to be tested. An open circuit will have infinite resistance. A voltmeter can be used to locate open circuits with the power turned on. The voltmeter will measure source voltage at all test points before the open circuit and zero volts at all points after the open circuit.

A fused jumper wire may be used to bypass a part of a circuit to locate a defect in the circuit. If wiring is suspect, a supply can be brought to a component with jumper wires to see whether it operates correctly. Besides supplying powering to motors and lights, jumper wires can bypass switches, relays, and solenoids. Jumper wires can also be used in some diagnostic routines to remove a component from a circuit to test for high resistance. A jumper wire should never be used to check circuits that are associated with computers, however. A high impedance voltmeter should be used for the latter.

7. Measure and diagnose the cause(s) of abnormal key-off battery drain (parasitic draw); determine needed repairs.

A parasitic draw occurs when a component stays on after it should shut off or when a component that should stay on begins to demand excessive current. Alarm systems, on-board diagnostics (OBD) II evaporative systems, power train control modules (PCMs), clock and radio memory, and convenience lighting are examples of systems that may stay active all the time or for quite some time after the vehicle is turned off. Light switches, relays, and generator voltage regulators are some very common causes of parasitic drain that can kill the battery over a few hours or a few days. Often the problems are intermittent and require good diagnostic skills to locate.

Many car manufacturers recommend measuring battery drain with a tester switch connected in series at the negative battery terminal. The drain test procedure must be followed in the vehicle manufacturer's service manual. A multi-meter with a milliamp scale is connected in parallel to the tester switch. When the tester switch is open, any current drain from the battery must flow through the multi-meter. Some computers require several minutes after the ignition switch is turned off before they enter sleep mode with a reduced current drain. Therefore, after the ignition switch is turned off and the tester switch is opened, wait for the specified time before recording the milliamp reading. The maximum specification is typically no more than 35 to 50 milliamps. Always consult each manufacturer specification if possible.

8. Inspect, test, and replace fusible links, circuit breakers, fuses, diodes, and current limiting devices.

When an ohmmeter is connected to a circuit breaker, fuse, or fuse link, the meter should read zero ohms if the component is working properly. An **open** circuit breaker, fuse, or fuse link causes an infinite ohmmeter reading. The current flow from an ohmmeter will not cause an automotive circuit breaker to open.

A test light is a good tool to use to quickly test fuses while connected to the vehicle. To perform this test, the ground clip should be connected to a good vehicle ground. The probe end of the test light then is used to touch the input and output side of each fuse. A good fuse will cause the light to illuminate on both sides of the fuse. A blown fuse will light up on only the input side. A fuse that will not light up on either side is not receiving electrical power and cannot be tested this way.

Diodes often are used in electric and electronic components. Many times these diodes cannot be serviced individually. If a diode fails, it is often necessary to replace the assembly. If a diode needs to be tested, a digital multi-meter (DMM) can be used. Turn the dial to diode test and then touch the test leads to the diode and note the reading on the meter. The test leads then should be reversed and then reconnected to the diode. A good diode will have a small voltage reading when connected one way and will have an out of limits (OL) reading when connected in reverse polarity.

9. Read and interpret electrical schematic diagrams and symbols.

All components have a special shape so they can be identified around the world on schematics. Some common components are shown in the following figure.

SYMBOLS USED IN WIRING DIAGRAMS			
+	Positive		Temperature switch
—	Negative		Diode
	Ground		Zener diode
	Fuse		Motor
	Circuit breaker	C101	Connector 101
	Condenser	→	Male connector
Ω	Ohms		Female connector
	Fixed value resistor		Splice
	Variable resistor	S101	Splice number
	Series resistors		Thermal element
	Coil		Multiple connectors
	Open contacts	88:88	Digital readout
	Closed contacts		Single filament bulb
	Closed switch		Dual filament bulb
	Open switch		Light-emitting diode
	Ganged switch (N.O.)		Thermistor
	Single pole double throw switch		PNP bipolar transistor
	Momentary contact switch		NPN bipolar transistor
	Pressure switch		Gauge

Typically, schematics have many common characteristics. See the following list for some helpful hints to follow when viewing wiring schematics:

- Power (B+) comes from the top of the page.
- Ground (Negative) comes from the bottom of the page.
- Wire colors are represented by letters next to the wires.
- Circuit numbers are represented by numbers next to the wires.
- Universal symbols for fuses, switches, relays, motors, resistors, and connectors are used to show the circuit layout.
- It is a good diagnostic practice to trace the flow of electricity in order to understand how the circuit works and then decide where to start the diagnosis.

See the following table for the common wire color codes.

Color	Abbreviations				
Aluminum	AL				
Black BLK	BK	B			
Blue (Dark)	BLU	DK	DB	DK	BLU
Blue (Light)	BLU	LT	LB	LT	BLU
Brown	BRN	BR	BN		
Glazed	GLZ	GL			

(Continued)

Color	Abbreviations				
Gray GRA	GR	G			
Green (Dark)	GRN	DK	DG	DK	GRN
Green (Light)	GRN	LT	LG	LT	GRN
Maroon MAR	M				
Natural NAT	N				
Orange	ORN	O	ORG		
Pink	PNK	PK	P		
Purple	PPL	PR			
Red	RED	R	RD		
Tan	TAN	T	TN		
Violet	VLT	V			
White	WHT	W	WH		
Yellow	YEL	Y	YL		

10. Diagnose failures in the data bus communications network; determine needed repairs.

Late-model vehicles typically have several on-board computers, which requires them to have a network that allows the computers to communicate with each other. This network also allows for scan tool communications. Typically, this network consists of two wires that are connected to the various modules and the data connector in parallel. The process of pulsing voltage to send signals is called multiplexing or bussing. The bus network wires are twisted continuously in order to help resist radio frequency interference (RFI) and electromagnetic interference (EMI) from entering this system. This method of sharing data among modules helps to eliminate redundant wiring to the modules that need the same sensor information.

Most late-model vehicles will have more than one data network on-board. These networks vary in the speed at which data is transmitted from one module to another. The fastest network is called CAN, which stands for Controller Area Network. This twisted pair of wires can transmit at speeds up to one million bits per second. Most high-speed networks use resistors at each end called terminating resistors, which are used to help reduce interference on the network. A technician can test to see whether these resistors are correct by measuring at the DLC pins 6 and 14 with the ignition switch turned off. The correct value of this reading should be 60 ohms.

Problems that can occur in the bus communication are opens, shorts, and unwanted resistance. Diagnosing these problems involves using the DMM in conjunction with a diagnostic trouble chart. The scan tool becomes less valuable when a communication problem occurs because it will not be able to receive data in the event of a wire failure. Another way to check for activity on the data bus network is to connect an oscilloscope to the network and monitor rapid voltage activity. If the voltage levels are changing continuously while the key is on, then it is probable that some communication is taking place on the data bus. It is always a good practice to verify that the scan tool will communicate with a similar vehicle before spending too much time attempting to diagnose the data bus. The DLC connector on the vehicle, as well as the scan tool wiring lead are potential locations of poor conductors.

Following is an example of a typical data bus network on a late-model vehicle.

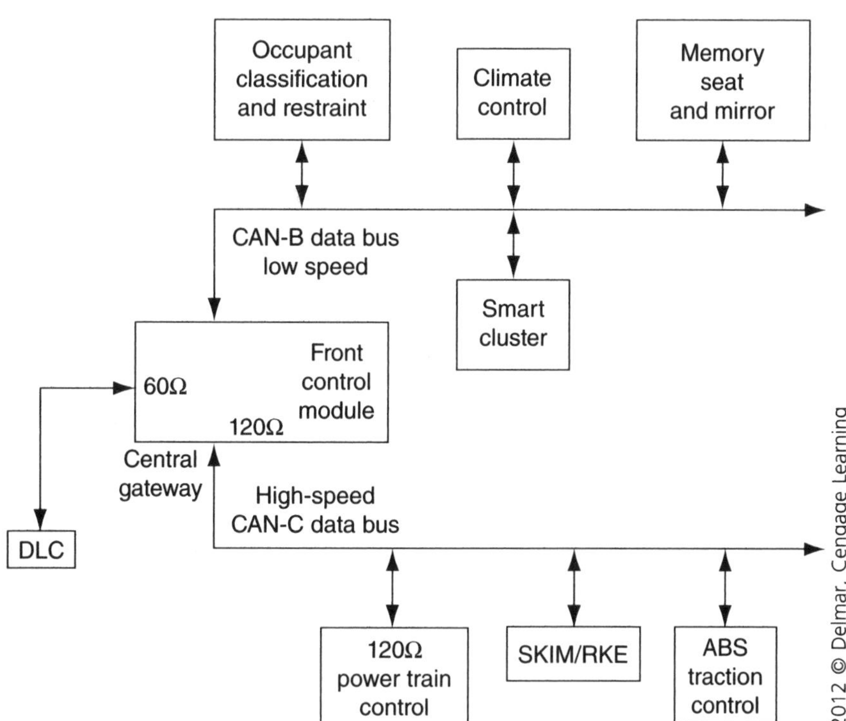

11. Remove and replace control modules; program as needed.

When it becomes necessary to replace any electronic control module (ECM), the technician must follow certain precautions. Prior to replacing any control module, the technician should perform careful diagnosis of the system in question. Items that need to be checked include the battery and charging system performance, connector quality and tightness, and the supply of good power and ground to the control module.

After verifying that a control module is the root problem and needs to be replaced, the technician should follow recommended procedures during its replacement. It is a good practice to remove the negative battery cable before disconnecting the control module. It is also recommended for the technician to touch a metal surface of the vehicle to discharge any electro static charge that may be present. Some manufacturers recommend wearing a grounding strap during a control module replacement.

The technician should make sure that all connectors are securely connected during the installation of a control module. It is also advisable to reinstall all of the hold-down fasteners for the module in order to prevent excess movement when the vehicle is operated.

Many control modules on late-model vehicles need to be programmed after the hardware is installed. The technician will need to have a programming tool or a scan tool that has a programming function to perform this action. In addition, a module programming database is needed to retrieve the correct program to upload to the new module. After programming the new module, it is advisable to operate the system that the module controls to assure correct operation.

B. Battery Diagnosis and Service (4 Questions)

1. Perform battery state-of-charge test; determine needed service.

When performing a battery state-of-charge test with a hydrometer, subtract 0.004 specific gravity points from the hydrometer reading for every 10°F (5.6°C) of electrolyte temperature below 80°F (26.7°C). During this test, 0.004 specific gravity points must be added to the hydrometer reading for every 10°F (5.6°C) of electrolyte temperature above 80°F (26.7°C). The maximum variation is 0.050 in cell-specific gravity readings. When all the cell readings exceed 1.265, the battery is fully charged.

The battery state-of-charge can be checked with a voltmeter as well. Following is a standard to go by:

12.6 volts = Fully charged

12.4 volts = 75 percent charged

12.2 volts = 50 percent charged

12 volts = 25 percent charged

A tool that is widely used to test batteries is the digital battery tester. This small tool measures the capacitance of the battery and gives feedback to the technician of the state-of-charge as well as if the battery needs to be charged or replaced. The following figure shows a digital battery tester.

2. Perform battery tests (load and capacitance); determine needed service.

The battery discharge rate for a capacity test is usually one-half of the cold cranking rating (CCR). Some batteries are also rated in "cranking amps," which is typically a higher number than the cold cranking amp (CCA) rating. A technician should be careful to not use the cranking amp rating when calculating the load-test amps. The battery must be at least 75 percent charged in order for this test to be valid. The battery is discharged at that rate for 15 seconds and the battery voltage must remain above 9.6V, with the battery temperature at 70°F (21.1°C) or above. The lower the temperature, the lower the voltage.

Most vehicle and battery manufacturers now recommend that batteries be tested with digital battery testers. These devices are small and accurate. The technician simply connects the leads of the tester to the battery and then adjusts the setting to the correct cold cranking amp setting. The tester quickly tests the capacitance of the battery. The test results will clearly reveal the state of the battery. The items that are determined with this test include the cold cranking amps, the state of charge, the battery voltage, and battery defects. Many of the testers are available with a small printer that can make a copy of each test. Most manufacturers require the test results be attached to the repair order on all battery warranty claims.

3. Maintain or restore electronic memory functions.

If battery voltage is disconnected from a computer, the adaptive memory in the computer is erased. In the case of a power train control module (PCM), disconnecting the power may cause erratic engine operation or erratic transmission shifting when the engine is restarted. After the vehicle is driven for about 20 miles (32 kilometers), the computer relearns the system, and normal operation is restored. If the vehicle is equipped with personalized items, such as memory seats or mirrors, the memory will be erased in the computer that controls these items. Radio station presets also will be erased. A 12 volt power supply from a dry cell battery may be connected through the power outlet connector to maintain voltage to the electrical system when the battery is disconnected.

4. Perform slow/fast battery charge in accordance with manufacturer's recommendations.

If the battery is charged in the vehicle, the battery cables should be disconnected during the charging procedure. The charging time depends on the battery state-of-charge and the battery capacity. If the battery temperature exceeds 125°F (51.7°C) while charging, the battery may be damaged. When fast charging a battery, reduce the charging rate when specific gravity reaches 1.225 to avoid excessive battery gassing. The battery is fully charged when the specific gravity increases to 1.265. Do not attempt to fast charge a cold battery.

5. Inspect, clean, and repair or replace battery(s), battery cables, connectors, clamps, hold downs, and vent tubes.

A battery may be cleaned with any number of battery cleaning products. The primary purpose of cleaning a battery is to eliminate surface discharge across the top of the battery and to remove any corrosion that has collected around the terminals. The electrolyte level should be checked and adjusted in batteries with removable caps. Maintenance-free batteries with built-in hydrometers indicate a low electrolyte level when the hydrometer is light yellow or clear. If the electrolyte level is low, the battery should be replaced. A low electrolyte level can be caused by a faulty voltage regulator that causes overcharging. When disconnecting battery cables, always disconnect the negative cable first. Replacement of the battery is straightforward, other than keeping in mind that cables and their connections must be clean to avoid voltage drops.

When removing battery clamps, take care not to stress the battery terminals. The clamps should be loosened first and then removed. If they do not come off readily, a commercially available puller should be used. To avoid battery damage, do not pry or apply sideways force to the terminals. Be careful to not overtighten side terminal connectors. The threads are in soft lead and can easily be stripped. Terminal contact surfaces should be cleaned with a suitable tool until the surfaces are clean and bright, thus ensuring good contact.

It is always wise to spray the cable clamps with a protective coating to prevent corrosion. Grease or petroleum jelly also will help prevent corrosion. Protective pads are available that go under the clamp and around the terminal to inhibit corrosion. If a cable has considerable corrosion, it might have corrosion that has worked up into the cable under the coating. When in doubt, do a voltage drop test on the cables and recommend replacement if 0.5 volts or higher is measured.

A final check of the battery should include making sure that the mounting brackets and hold-down components are in place. A battery that moves around while the vehicle is in motion is less likely to hold up as well as it should.

6. Jump start a vehicle using jumper cables and a booster battery or auxiliary power supply.

The accessories must be off in both vehicles during the boost procedure. The negative booster cable must be connected to an engine ground in the vehicle being boosted. Always connect the positive booster cable followed by the negative booster cable and complete the negative cable connection last on the vehicle being boosted. Do not allow vehicles to contact each other. When disconnecting the booster cables, remove the negative booster cable first on the vehicle being boosted.

7. Perform starter current draw test; determine needed repairs.

A starter draw test is performed with a specialized ammeter that is generally also used for testing generators. An inductive clamp is placed on the cable to the starter. Most inductive pick-ups are polarized, which means that the pick-up must be clamped around the cable in the proper direction. Look for an arrow on the pick-up. The arrow always points to the device being tested. Disable the ignition system and either have an assistant crank the vehicle or use a remote starter button. There will be a slight spike in the current as the starter begins to crank. In most applications, a starter will demand 130–200 amps, depending on the design of the starter and the size of the engine. Gear reduction starters do not take as much current to operate.

High starter current draw, low cranking speed, and low cranking voltage usually indicate a defective starter. This condition also may be caused by internal engine problems. Low current draw, low cranking speed, and high cranking voltage indicate excessive resistance in the starter circuit.

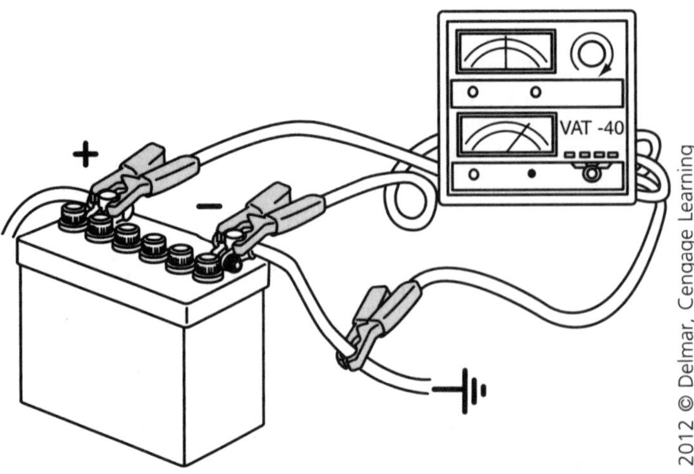

8. Perform starter circuit voltage drop tests; determine needed repairs.

Measure the voltage drop across each component in the starter circuit to check the resistance in that part of the circuit. The ignition and fuel system must be disabled while making these tests. Read the voltage drop across each component while the starting motor is operating. For example, connect the voltmeter leads to the positive battery terminal and the positive cable on the starter solenoid and crank the engine to measure the voltage drop across the positive battery cable. A general specification for this test is any component reading less than 0.5 volts is acceptable. Generally a total of the voltage drops should not exceed 1.5 volts. Refer to the particular manufacturer for exact specifications.

9. Inspect, test, and repair or replace starter, relays, solenoids, modules, switches, connectors, and wires of starter circuits.

Relays and switches in the starting motor circuit may be tested with an ohmmeter. When an ohmmeter is connected across the relay or switch contacts, the meter should provide an infinite reading if the contacts are open. If the relay or switch contacts are closed, the ohmmeter reading should be very low. When the ohmmeter leads are connected across the terminals that are connected to the relay winding, the meter should indicate the specified resistance. A resistance below the specified value indicates a shorted winding, whereas an infinite reading proves that the winding is open. In high amperage circuits, more than a 0.5 volt drop may indicate a problem. Relays often develop corrosion on contacts, which can cause problems with either power feed to the controlled device/ system or activating the relay. Many relays have diodes or resistors in parallel with the coil to suppress voltage spikes when the relay is de-energized. Starter solenoids are examples of large relays controlled by a fairly low amperage ignition switch.

The figure below is an example of a common starting circuit.

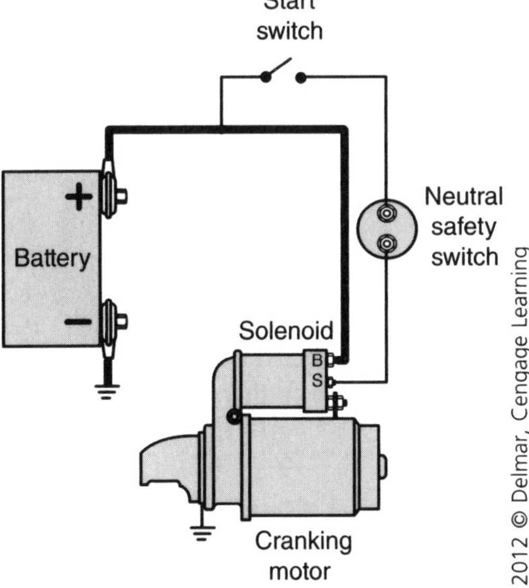

The ohmmeter leads must be connected across the "s" terminal and the "motor" terminal to test the pull-in winding of the starter solenoid. Connect the ohmmeter leads from the "s" terminal to the ground to test the hold-in winding of the starter solenoid. Check starter relays and solenoids for excessive resistance in the high-current contacts. A voltage drop across the load side of these devices while in operation is the best method to test for burnt contacts.

The negative battery cable should be removed before starting the procedure of removing the starter. Some starters use alignment shims. The purpose of adding shims between the starter and its mounting surface is to adjust the depth of the starter drive gear when it engages the flywheel. Virtually all vehicles have precision-engineered starter mounts that avoid the need for this.

The replacement starter should be closely inspected prior to installation to make sure that it is the correct model. It is also a good practice to bench test the replacement starter to make sure it is functional prior to installing it on the engine. The starter wires and connections should be inspected for damage, corrosion, and broken insulation during this procedure.

10. Differentiate between electrical and engine mechanical problems that cause a slow crank, no crank, extended cranking, or a cranking noise condition.

The first step to checking a starting problem should be to check the battery condition. Many starting problems can be attributed to battery problems. Do a complete series of battery tests to confirm the condition of the battery. Perform a visual inspection of the starting system to find loose connectors and frayed wires. After the visual inspection, if the engine will not turn over, try to turn the engine over manually by turning the crankshaft bolt. If the engine will not turn with this method, then internal engine problems are present and need to be located. Examples of engine conditions that could cause this condition include fluid in the cylinders, damaged engine bearings, or a locked-up engine.

Slow cranking also can be caused by engine problems. An engine that is not timed correctly will cause slow cranking as well as increased starter current draw. It may be necessary to perform an engine compression test and visually inspect the timing marks in order to determine whether it is timed correctly.

The ways to determine whether the starter is causing the no crank or slow crank include checking the starter current draw as well as checking for excessive voltage drops in all of the starter and battery components. See Tasks B.7 and B.8 for further explanation.

An extended cranking problem is usually caused by the fuel or ignition system. If the battery voltage and cranking amperage are acceptable, then the starter should create the correct cranking speed to allow the engine to start. If these variables are within the correct range, then the fuel and ignition system should be checked for problems.

Items that could cause unusual cranking noises include a worn starter drive gear, a worn ring gear, loose starter mounting bolts, and incorrect drive gear to ring gear clearance. A close inspection of these items will typically reveal the location of the noise. The mounting bolts should be at the correct torque. The starter drive gear and the ring gear should not be worn excessively. Excessive wear will leave the surfaces shiny and misshaped. If incorrect clearance is suspected, a technician can use gear marking compound on the ring gear and then briefly operate the starter. An inspection will reveal the contact point of the gears. Shims can be removed if the clearance is too wide or added if the clearance is too close.

C. Charging System Diagnosis and Repair (5 Questions)

> *Note:* In 1996 SAE J1930 terminology was adapted to standardized component identification. This standard adopted the name "generator" to refer to the component commonly known as an "alternator." Both terms are used interchangeably in the ASE tests.

1. Diagnose charging system problems that cause a no-charge, a low charge, or an overcharge condition; determine needed repairs.

The table gives some examples of various charging problems that a technician might face. Refer to the other tasks in this section for procedures to test the components and remember that many of these problems are circuit problems that can be traced by performing voltage drops on loaded circuits.

Condition	Possible Cause
Undercharging	• Loose generator belt
	• Faulty voltage regulator
	• Faulty generator
	• Poor connections at generator or battery
	• Faulty battery
No Charging	• Loose or missing generator belt
	• Faulty voltage regulator
	• Faulty generator
	• Poor connections at generator or battery
	• Faulty battery
Overcharging	• Faulty voltage regulator
	• Faulty generator
	• Poor connections at generator or battery
	• Faulty battery

2. Inspect and reinstall/replace pulleys, tensioners, and drive belts; adjust belts and check alignment.

An undercharged battery may be caused by a slipping generator belt. A slipping belt may be caused by insufficient belt tension or a worn, glazed, or oil-soaked belt. Belt tension may be tested with a belt tension gauge or by measuring the belt deflection in the center of the belt span. A belt should have 0.5 in (12.7 mm) of deflection for every foot (30.5 cm) of free span. Many serpentine belts have an automatic spring-loaded tensioner with a belt wear scale.

3. Perform charging system voltage output test; determine needed repairs.

An important test that can be performed on a charging system is the voltage output test. Charging system voltage typically is tested with a digital voltmeter. The positive voltmeter lead connects to the battery positive terminal, and the negative voltmeter lead connects to the battery ground terminal. It is important to remember that the battery needs to be at least 75 percent charged in order to check generator output voltage accurately. Charging output voltage is typically 13.5 volts to 14.5 volts. If the output tests show the voltage lower than these readings, then the alternator could be weak, and further tests will need to be performed. Readings above 14.5 volts are also a concern.

The voltage regulator should be closely checked if these levels are found. Some generators utilize an internal voltage regulator that cannot be checked separately from the generator itself.

4. Perform charging system current output test; determine needed repairs.

Most charging system tests are done with special charging system testers that have a device called a carbon pile that allows the technician to create a simulated load on the generator. Before beginning the test, the technician must determine what the rated output of the charging system is. After that is determined, the technician can perform the tests by applying a load to the generator with all of the other vehicle accessories turned off. If the system cannot generate the specified output, the technician must determine whether the cause is related to wiring issues or faults with the generator or voltage regulator. Each manufacturer has tests designed to reach this end, but a wiring diagram and digital multimeter (DMM) can be very useful diagnostic tools.

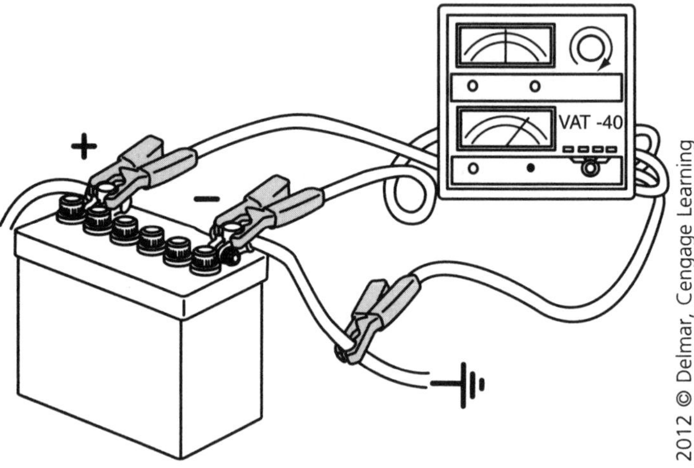

5. Inspect and test generator (alternator) control components, including computers/regulators; determine needed repairs.

The voltage regulator controls the field circuit in the generator to keep the output voltage at about 13.5 to 14.5 volts, depending on the electrical requirements. In some applications, the voltage regulator is external of the generator and requires testing of wiring circuits before determining that there is a problem with the regulator. In most cases if the regulator has failed, the generator will have sustained some damage as well. With modern generators creating as much as 150 amps, internal damage can occur quickly if the regulator fails. Before replacing a regulator, carefully inspect the generator and determine its age to avoid a comeback when the generator fails, often taking the new regulator with it.

Many late-model vehicles have incorporated the voltage regulator into the engine computer. This method of voltage regulation works very well because the engine computer is monitoring many different sensors, such as battery temperature, ambient air temperature, and engine speed, as well as charging voltage. Having all of this data allows the engine computer to match charging output with the many variables that the vehicle is operating in. The technician also has the ability to use a scan tool in the diagnosis of these charging systems. This allows the technician to view data and retrieve trouble codes that can assist the diagnosis process. In addition, some scan tools allow the technician to control the charging output with a function test.

6. Perform charging circuit voltage drop tests; determine needed repairs.

A voltmeter may be connected from the generator battery wire to the positive battery terminal to measure voltage drop in the charging circuit. Many car manufacturers recommend a 10-amp charging rate while measuring this voltage drop. When the voltage drop is more than specified, the circuit resistance is excessive. High-charging circuit resistance between the generator battery terminal and the positive battery terminal may cause an undercharged battery and an overworked generator. The negative side of the charging circuit should also be checked. The voltmeter leads should be connected to the case of the alternator and the battery negative terminal with the engine running. A reading of 0.2 volts or higher indicate a problem. Possible causes could be broken brackets, loose generator mounting bolts, or a bad engine-to-body grounding strap.

7. Inspect and repair, or replace connectors and wire of charging system circuits.

Always use rosin core solder when repairing electrical circuits by soldering. Always insulate bare wires with heat shrink tubing. If the wire contains a shield wire, solder the drain wire separately. When using a heat and crimp seal, apply heat until the sealant appears at both ends. Crimp and seal connectors are suitable if care is taken to strip the proper amount of insulation from the wires. This will ensure adequate engagement in the connector without exposing the bare conductor. These connectors are suitable for charging system repairs because they can be heated to melt the ends, which prevents water from entering the connector.

8. Remove, inspect, and replace generator (alternator).

Disconnect the negative battery terminal before removing the generator. Never use air-powered tools when removing or installing the small fasteners used to attach the charging wires. This will reduce the chance of over-tightening the fasteners and damaging the internal generator connections. The replacement generator should be inspected closely to ensure it is the correct model. The drive pulley should be the same diameter as well as have the same number of pulley grooves as the old generator.

D. Lighting Systems Diagnosis and Repair (6 Questions)

1. Diagnose the cause of brighter than normal, intermittent, dim, continuous, or no operation of exterior lighting; determine needed repairs.

Formerly one of the most basic circuits in a vehicle, the headlights have automatic dimming, light-sensing switches to automatically turn them on, with auxiliary driving lights integrated into them. Keep in mind that these items have their own controllers and relays that operate them. A DMM and wiring diagrams often will be used to isolate problems. Ohm's Law applies to every lighting circuit, too.

Concern	Possible Cause
Brighter than Normal	• High-charging system voltage • Dimmer switch in high beam position • Flash to pass or high beam relay stuck on • Bulb problems • Incorrect setting of automatic dimming sensor
Intermittent Operation	• Connection issues with switches, firewall bulkhead connectors, in-line connectors, head lamp connectors, fuse, or relay blocks • Weak circuit breakers • Problems with head lamp relays • Shorts causing circuit breaker shutdown • Bulb problems • Faulty relays • Faulty switches (headlight or dimmer)
Dim	• Poor grounds • Contact corrosion in head lamp relays • High resistance in connections at switches, firewall bulkhead connectors, in-line connectors, head lamp connectors, fuse, or relay blocks • Bulb problems
No Headlights	• Bulb problems • Open or high-resistance connection issues with switches, firewall bulkhead connectors, in-line connectors, head lamp connectors, fuse, or relay blocks • Damaged circuit breakers or fusible links • Faulty relays • Faulty switches (headlight or dimmer) • Shorts causing circuit breaker shutdown

Daytime running lights (DRL) are typically part of the vehicle's high-beam circuit. The control circuit is connected directly to the vehicle's ignition switch so the lights are turned on whenever the vehicle is running. The circuit is equipped with a module that reduces battery voltage to approximately 6 volts. This voltage reduction allows the high beams to burn with less intensity and prolongs the life of the bulbs. When the headlight switch is moved to the "on" position, the module is deactivated and the lights work normally.

Diagnosis of these systems should begin by identifying whether the problem is in the DRL system or the headlight system. If the problem is in the headlight system, service to the circuit and lamps is conducted in the same way as for vehicles that are not equipped with DRL. If the problem is in the DRL system and the headlights work normally, only the part of the circuit that is unique to the DRL can be the problem.

2. Inspect, replace, and aim/level headlights/bulbs including high-intensity discharge (HID) systems and auxiliary lights (fog lights/driving lights).

When servicing the halogen or xenon bulbs used in head lamps and auxiliary lighting, always turn off the lights and allow the bulbs to cool. In addition, keep moisture away from the bulb, handle the bulb only by its base, do not scratch or drop the bulb, and coat the terminals of the bulb or the connector with dielectric grease to minimize corrosion. Sealed beam and standard bulbs should be placed securely into their retaining plates or fixtures. It should be noted that replaced HID lamp assemblies likely will be brighter than the original lamps due to their age. The owner of the vehicle should be counseled about this to see whether they would like to replace both sides at once.

There are various way to set the correct aim of vehicle headlights. Aiming equipment is sometimes used to align the headlights to the correct settings. Some vehicles are built with headlight adjustment mechanisms built into the headlamp housing.

When diagnosing auxiliary lighting systems, it is important to keep in mind that there are many different strategies for their function. Some will work only when the low beams are on, and high beam settings may turn them off. The manufacturer's information and a wiring diagram are critical tools to the successful solution of the customer's concern.

3. Inspect, test, and repair or replace switches, relays, bulbs, LEDs, sockets, connectors, wires, and controllers of exterior lighting.

The components of a typical headlight system of late-model vehicles include the headlight switch, the dimmer switch, the headlight relay, wires, connectors, lamp sockets, and bulbs. Diagnosis of these components is similar to other electrical components. A good check of the fundamentals should be performed. This check would include visually checking the whole system for any physical problems or damage. A quick voltage test of the fuses, bulbs, switches, and relays would soon follow.

Some headlight systems are controlled by logic devices such as body control modules or lighting control modules. These systems typically can be diagnosed with a scan tool. A scan tool interfaces with these computers to monitor data and trouble codes as well as command output tests of the electrical items controlled.

The following figure shows a schematic of a simple headlight circuit.

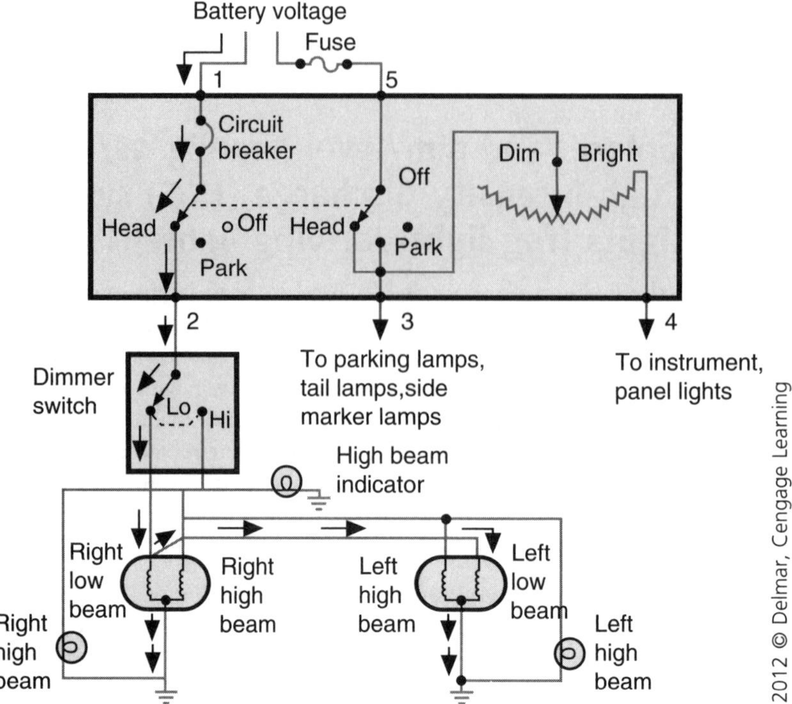

If the correct voltage is supplied and dropped at various lights on the vehicle, then the problem has to be in the bulb. Care should be taken when replacing the bulb to not get any fingerprints on the bulbs. This will shorten the bulb life and cause customer dissatisfaction. All connector and wiring repairs should be made using water-resistant methods such as using solder and heat shrink or crimp-and-seal connectors.

4. Diagnose the cause of turn signal and/or hazard light system malfunctions; determine needed repairs.

Most turn signal flashers change the flash rate when the electrical current changes. This design provides feedback to the driver when a bulb burns out. A few manufacturers use electronic flashers; flasher relays will not give the driver feedback on the instrument panel that a lamp is burned out.

The hazard flasher is typically a heavy-duty flasher. Some vehicles will have a trailer wiring harness installed that allows the vehicle to connect to a trailer lighting system. It is common for the turn signal flash rate to change when the trailer is connected. A heavy-duty flasher can be installed in place of the turn signal flasher to eliminate this problem.

Some common problems that occur in the turn and hazard lamp circuits include shorted bulbs, incorrect bulbs, bad grounds, and faulty flashers. In many stoplight circuits, voltage is supplied to the brake light switch from the fuse box. When the brakes are applied, brake pedal movement closes the stoplight switch. This action supplies voltage to the stoplights and the center high-mounted stoplight. In many stoplight systems, the stoplight filaments share the same light bulb encasement as the taillights and turn signals.

A logical diagnostic approach should be used when diagnosing turn signal and hazard light system problems.

With the addition of the neutral safety or back-up light switch, everything we have discussed in other areas of the lighting is true of back-up lamps as well. To review and expand, intermittent problems typically are caused by loose or poor connections. All connections must be clean and strong in order for all parts of a circuit to operate normally. Dim or poor operation of a lamp or other component is caused by high resistance in the circuit. The most common cause for high resistance is corrosion. When a controller or switch fails to turn on or turn off a component, the switch is bad, or there is a short in the control circuit or an open somewhere in the circuit. A quick voltage check at the component should help determine the cause. After the cause is identified, the type of testing needed to further define the problem will be known.

When the ignition switch is on and the back-up light switch is closed, voltage is supplied through these switches to the back-up lights. The gear selector linkage operates the back-up light switch. This switch might be mounted on the steering column, shifter, or transmission.

5. Inspect, test, and repair or replace switches, flasher units, bulbs, sockets, connectors, wires, and controllers of turn signal and hazard light circuits.

Turn signal switches have been incorporated into a device called a multifunction switch. This multifunction switch typically is located on the steering column and can be serviced by taking the plastic housing that surrounds the column off. The turn and hazard flashers typically are located under the dash area near the fuse panel. Replacing these components is a fairly simple process. Flashers typically just plug into a connector or fuse panel.

Professional repair methods should be followed when servicing the electrical components of the turn and hazard lamp systems. All wire repairs should be performed with either solder or crimp-and-seal connectors. Dielectric grease should be used when servicing bulbs and socket for these circuits.

Many back-up light switches are mounted on top of the steering column under the dash, and these switches often are combined with the neutral safety switch. The combination back-up and neutral safety switch is operated by the gearshift tube in the steering column. The back-up light switch might be adjusted by loosening the mounting bolts and rotating the switch. Others are incorporated into the PRNDL switch mounted at the transmission. Some manufacturers incorporate the back-up lamps operation into the PCM or TCM. These systems typically require the use of a scan tool to test and diagnose the operation of the back-up lights.

6. Diagnose the cause of intermittent, dim, continuous, or no operation of courtesy lights (dome, map, vanity, cargo, trunk, and hood); determine needed repairs.

Interior lights in late-model vehicles can be operated in two different ways. One way to operate the lights is with simple switches the directly control power or ground to the lights. These switches can be tested with a voltmeter the same way that any electrical switch is tested. A second way that many late-model vehicles control the interior lighting is with the body control module (BCM). In these circuits, the switches can be configured as "open/closed" switches or they can be resistive switches that vary their resistance as they operate. The BCM style of interior lighting can be diagnosed with a scan tool connected to the vehicle. The scan tool will display the output of the resistive switches to reveal the functionality of the switch. Both styles of interior lighting systems need to have quality connections and wire repairs performed when problems arise.

7. Inspect, test, and repair or replace switches, relays, bulbs, sockets, connectors, wires, and controllers of interior lighting circuits (courtesy, dome, map, vanity, cargo, trunk, and hood).

Courtesy lights on late-model vehicles typically are included in the body control module (BCM) operation. This allows the lights to be slowly phased in and out when the doors are opened and closed. Many of these systems still use the old-style switches that are one- or two-wire design, but instead of being directly wired into the circuit, they are used as inputs to the BCM. Since these lights are now included in the logic circuits on the vehicle, it is less likely to run the battery down if one gets left on. Many of these vehicles will sense a load on the battery when the vehicle has sat for a period of time and turn off the lights that are staying on, which prevents a discharged battery.

These systems still use incandescent bulbs so they do wear out after years of operation. These systems also are protected by fuses that will blow if a short circuit occurs. It is possible that the door switches can get out of adjustment from time to time. If the lights do not go off and stay off with the doors closed, then the adjustment should be checked.

8. Inspect, test and repair or replace trailer wiring harness, relays, connectors, and controllers.

Many trucks and sport utility vehicles are equipped with trailer wiring connectors that can be used to supply vehicle lighting to a trailer that is being pulled. Most vehicles of this category will come from the factory with this option. If the vehicle is not equipped from the factory, then kits can be added to the vehicle to accomplish this task. It is wise to use the kits of the plug-and-play design. These kits include a separate controller and do not require any splicing and potential damaging of the existing lighting circuits. These harnesses and connectors can be tested with a 12 volt test light. Voltage for stop, turn, and running lights should be available on these circuits.

E. Instrument Cluster and Driver Information Systems Diagnosis and Repair (6 Questions)

1. Diagnose the cause of intermittent, dim, no lights, continuous operation or no brightness control of instrument lighting circuits.

Concern	Possible Cause
Brighter than Normal	• High-charging system voltage • Bulb problems
Intermittent Operation	• Connection issues with switches, firewall bulkhead connectors, in-line connectors, lamp connectors, fuse, or relay blocks • Problems with lamp relays • Shorts causing circuit breaker shutdown • Bulb problems • Faulty relays

Concern	Possible Cause
	• Faulty switches (headlight or auxiliary) • Poor grounds
Dim	• Poor grounds • Contact corrosion in lamp relays/switches • High resistance in connections at switches, firewall bulkhead connectors, in-line connectors, head lamp connectors, fuse, or relay blocks • Bulb problems
No Lights	• Bulb problems • Open or high-resistance connection issues with switches, firewall bulkhead connectors, in-line connectors, head lamp connectors, fuse, or relay blocks • Damaged circuit breakers or fusible links • Faulty relays • High beam switch in the on position or faulty (fog/driving lights) • Faulty switches (headlight or dimmer) • Shorts causing circuit breaker shutdown

Be sure to take note of whether a problem is system wide or associated with only one bulb. If a problem is limited to one lamp, the diagnostic approach is different than if it affects all of the lights in the circuit (or the body panel). In situations in which only one is affected, the technician should be looking for something that can affect only the one such as wires, bulbs, or a ground. In a circuit-wide problem, the technician should be looking for the common areas. If all lamps are out in a tail lamp, the problem could be a common ground connection.

2. Inspect, test, and repair or replace switches, relays, bulbs, LEDs, sockets, connectors, wires, and controllers of instrument lighting circuits.

Regardless of the light circuit being inspected, check some common things to identify the cause of operational problems. Visually inspect the entire light circuit for loose and/or corroded connections. Likewise, also inspect the bulb terminals.

Sometimes the cause of a light problem may be the lamp itself. It is safe to assume that when one bulb is not working, the bulb is bad. The problem can be verified by replacing the bulb with a new one. When doing this, make sure to install the same type of bulb as the vehicle originally was equipped with and make sure that it is properly installed. If more than one bulb in a circuit is not working, it is very likely that the cause of the problem is not the bulbs. Check the system for an open. Light bulbs can be checked with an ohmmeter. If the meter reading across the terminals is infinite, an open (burned-out) bulb is indicated. Some computerized instrument panels use a logic device to control the instrument lighting. These systems typically can be diagnosed by using a scan tool that communicates with the instrument panel. The scan tool can be used to retrieve data and trouble codes from the instrument panel.

3. Diagnose the cause of intermittent, dim, no lights, continuous operation, or no brightness control of instrument lighting circuits; determine needed repairs.

When troubleshooting instrument lighting problems, the technician should inspect whether a problem is system wide or associated with only one bulb. If a problem is limited to one lamp, the diagnostic approach is different from one that affects all of the lights in the circuit or the body panel. For situations in which only one lamp is affected, the technician should look for a cause that would affect only the one lamp, such as wires, bulbs, or a ground. In a circuit-wide problem, the technician should be looking for the common areas. If all lamps are out in a tail lamp, the problem could be a common ground connection.

A rheostat is connected in series with the instrument cluster bulbs. This rheostat is operated by the headlight switch knob or by a separate control knob. When the rheostat control knob is rotated, the voltage to the instrument cluster bulbs is reduced. This action lowers the current flow and reduces the brilliance of the bulbs. The instrument cluster bulbs are connected in parallel to the battery. If one bulb burns out, the other bulbs remain illuminated.

4. Diagnose the cause of, high, low, intermittent, or no readings on electronic instrument cluster gauges; determine needed repairs.

> *Note:* Diagnosing causes of abnormal charging system gauge readings or warning lights limited to dash units and their electrical connections; other causes of abnormal charging system indications are covered in category C.

Many vehicles are equipped with thermal-electric gauges. These gauges contain a bimetallic strip surrounded by a heating coil. The pivoted gauge pointer is connected to the bimetallic strip. The sending unit contains a variable resistor. In a fuel gauge, this variable resistor is connected to a float in the fuel tank. Many late-model instrument panels incorporate electronic gauges that are driven by integrated circuits that take data lines from the PCM and convert them into analog gauge readings. In normal service, these units are serviced by specialists and not by technicians.

All gauge inputs come from some kind of sending unit. These are the most common failures and causes of inaccurate gauge readings. Since these units are almost all variable resistors, anything that alters resistance will cause incorrect gauge readings. Gauges that typically work this way are temperature, fuel, and oil pressure gauges.

Tachometers and electronic speedometers are examples of units that run off of a pulse, frequency, or voltage. Tachometers are driven by the negative side of the primary ignition and are calibrated for the number of cylinders the engine has. Connections are usually the only problems that will occur. Speedometers have probably got the highest incidence of failure along with fuel and oil pressure gauges. Electronic speedometers are driven by a pulsed signal that is almost always an AC sine wave that comes from a signal generator attached to the transmission or differential. The speedometer generally is calibrated to expect a certain number of pulses per mile or is calibrated to the tire, transmission, and gear-ratio package of the vehicle. This information may be managed by the PCM and translated for the instrument panel speedometer.

Many electronic instrument displays provide an initial illumination of all segments when the ignition switch is turned on. This illumination proves the operation of the display segments. During this initial display, all the segments in the electronic instrument displays should be brightly illuminated for a few seconds. If some of the segments are not illuminated, replace the electronic instrument cluster. When none of the segments is illuminated, check the fuses and voltage supply to the display. Many electronic instrument displays have self-diagnostic capabilities. In some electronic instrument displays, a specific gauge illumination or digital display will indicate defects in the display. Many late-model electronic displays should be diagnosed with a scan tool. The scan tool can view live data and trouble codes, as well as perform output tests to allow the technician to command various outputs from the instrument cluster.

5. Diagnose the cause of constant, intermittent, or no operation of warning lights, indicator lights, audible warning devices, and other driver information systems; determine needed repairs.

One of the causes for constant operation of a warning or indicator light could be the control circuit is shorted to ground. Another cause for a light that stays on could be a switch or sender shorted out. If the warning light works intermittently, the most likely cause is a loose wire or connector causing an occasional short. If the light does not work at all, the bulb is burned out or there is an open in the circuit.

Many warning lights are operated by the body control (BCM) control module. The various switches and sensors send signals to the BCM as they change their state. When one of these signals is received, the BCM grounds the appropriate circuit.

Performing service on the bulbs, sockets, connectors, and wires is the same as for all of the other light systems on the vehicle. Solder and heat shrink as well as crimp-and-seal connectors can be used to repair the wires in these circuits.

Most late-model vehicles that use a body control module use the inputs from door switches along with sensors in the ignition switch and seat belts to execute warning strategies. An example is if the key is in the ignition and the door is open with the engine off: A chime will sound. Another example is when the vehicle is in motion or the transmission is in gear and the passenger seat does not have a complete seat belt circuit: The warning lamp on the dash will light and a chime will sound.

The body control units have many inputs, including, seat belts, door switches, wiper switch, horn, anti-theft inputs, vehicle speed inputs, headlight switch, brake switch, and gauge warnings, to mention some of the most common. They use a series of outputs that include warning lights on the dash, different chimes and buzzers, to let the driver know what is happening or to accomplish automatic functions, such as locking the doors based on vehicle speed or gear shift input. When diagnosing them, a scanner is almost always the quickest route. Beyond that, pinpoint tests can be performed with ohmmeters and voltmeters. Everything that applies to any other switch or output device is true here.

Various types of tone generators, including buzzers, chimes, and voice synthesizers, are used to remind drivers of a number of vehicle conditions. These tone generators should be checked for operation by running each through the prescribed self-test mode. Audible warning devices are generally activated by the closing of a switch. A tone is emitted to warn the driver that something in the system is not functioning properly or that a situation exists that must be corrected. Many of the warning systems on today's vehicles are triggered by a PCM or BCM and may be integrated into the unit. Always refer to the testing methods recommended by the manufacturer when testing these systems.

6. Inspect, test, and repair or replace bulbs, sockets, connectors, switches, relays, sensors, timers, wires, gauges, sending units, electronic components, and controllers of electronic instrument clusters and driver information system circuits.

Circuits for electronic instrumentation are very much like those found in a conventional instrument circuit. When testing electronic instrument components, it is important to remember that electronic gauges can be either analog or digital. Analog gauges give the ability to show a constant change in value. Digital circuits operate in one of two states: on or off. The pulsing of the circuit (on and off) is what determines the readings on the instruments.

In order for an electronic gauge to display an accurate reading, it must receive an accurate signal from its sensor or sensors. The diagnosis should include checking the connection and wiring at the sending unit. The gauge should respond when the sending unit is unplugged or grounded out.

A typical gauge circuit is no more than a simple series circuit with a variable resistor. The variable resistor or sending unit responds to the changes in fluid level or operating condition of the engine. Because the gauge is part of this series circuit, a change in circuit resistance will cause a change in voltage to the gauge and in current flow through the circuit. Unwanted resistance, from corroded terminals or similar problems, will cause the gauge to read incorrectly, as will high operating voltages and defective sending units. Sending units can be tested with an ohmmeter. Specifications are normally given for the sending unit in a variety of positions or conditions. Voltage checks at the gauge may also be necessary to diagnose the gauge circuit. Sending units, gauges, controllers, and printed circuit boards are replaced, not repaired, when they are found to be defective.

Many late-model vehicles are equipped with driver information systems that display various types of data on a screen. These systems operate by being integrated with the communication network on the vehicle and then displaying the necessary data onto a screen for the driver to view. A scan tool is needed to diagnose problems in this system. If a display system is not functioning correctly, the technician should check the connections for tightness. If no problems are found with the connections, then the driver information display assembly can be replaced as a unit.

F. Body (11 questions)

1. Diagnose operation of comfort and convenience accessories and related circuits (such as: power window, power seats, pedal height, power locks, truck locks, remote start, moon roof, sun roof, sun shade, remote keyless entry, voice activation, steering wheel controls, backup camera, park assist, and auto dimming headlamps); determine needed repairs.

Many late-model vehicles are equipped with dozens of comfort and convenience accessories. Accessories such as power windows, power seats, power locks, trunk locks, pedal height motors, sun roofs, and moon roofs use bi-directional motors. These systems operate in a similar way because most of them use reversible motors that are controlled by momentary contact switches. The motors change direction when the electrical polarity is reversed. Both power (B+) and ground is typically supplied at the main switch for these systems. If either power or ground is lost at the main switch, then all operation will stop. If these systems are controlled by a control module, then a scan tool can be used to assist the technician in the troubleshooting process.

Systems such as remote keyless entry, voice-activated controls, backup camera, park assist and auto dimming headlights will always be controlled by a control module. As with any computer system, these systems use various input sensors that send signals to the logic device that is then processed to possibly create an output action. Scan tools are required in order to diagnose these systems when problems occur. Digital multi-meters (DMMs) will also be necessary to test the electronic circuits of these very advanced systems.

2. Inspect, test, and repair or replace components, connectors and wiring of comfort and convenience accessories.

Technicians who repair the advanced electronic systems on late-model vehicles will need to practice high degrees of professionalism due to their complexity. It is important to carefully perform the diagnosis of all electrical and electronic systems. Performing a thorough visual inspection will sometimes uncover the problems that are causing a malfunction. All connections should be secure and locked. Wiring harnesses should be correctly routed and supported by wire brackets. A misrouted wire harness can lead to potential electrical failures due to rubbing against a piece of metal or wires stretched beyond their normal length. Professional wire repair techniques should also be used on these systems to prevent dirt and moisture from entering the circuits.

Many manufacturers recommend disconnecting the battery cable before servicing any control module. Technicians should also take precautions regarding static electricity before handling any control modules. This can be done by wearing a grounding strap or by touching a metallic surface on the vehicle before handling the module.

3. Diagnose operation of heated and cooled accessories and related circuits (such as: heated/cooled seats, heated steering wheel, heated mirror, heated glass, and heated/cooled cup holders); determine needed repairs.

Many late-model vehicles are equipped with various heated and cooled systems to add comfort, safety, and convenience for the driver and passengers. The heated systems typically use thermo-electric strips that create heat when power is supplied to them. These heated systems require a large amount of current, so power is supplied them by a relay. A control module directs the relay when the heated system is requested by the driver or passenger. These heated systems can be diagnosed by using good diagnostic strategy.

Some late-model vehicles have cooled seats. Many of these systems include perforated seats which allow air from the A/C system to be routed to the seat cushions. Manufacturers often install a blower motor at the bottom of the seat to assist in moving cooled air through the seat cover. The seats on these vehicles should be routinely cleaned to prevent the perforated holes from becoming blocked. The blower motor can be activated with a scan tool to assist in troubleshooting.

Some vehicles have heated and cooled cup holders. Heating and cooling of the cup holders is accomplished by using electric grids that can add heat or remove heat, depending on the polarity of the current. The driver and passenger can choose which temperature is delivered by adjusting a switch.

Heated steering wheels operate through a heater grid built into the steering wheel. The current path travels through the clock spring/ribbon which allows energy to be delivered to a rotating steering wheel. A scan tool can be used to communicate with the control module when diagnosing problems.

4. Inspect, test, and repair or replace components, connectors, and wiring of heated and cooled accessories.

Repairing heated and cooled accessory systems requires an understanding of how each system functions. It is advisable to determine if the system to be repaired is controlled by a logic device. Scan tools can be used to perform diagnostic routines such as retrieving codes and viewing live data. Like all of the other electronic systems on the vehicle, professional wiring repair techniques should be followed on heated and cooled circuits.

5. Diagnose operation of security/anti-theft systems and related circuits (such as: theft deterrent, door locks, remote keyless entry, remote start, and starter/fuel disable); determine needed repairs.

Modern factory installed anti-theft systems use a dedicated control module or design the functions into an existing module. Most factory installed systems are a passive only system, which means the system is designed only to start the car when the correct key is inserted into the ignition. The system will disable the starter system, fuel delivery system, ignition system, or any or all of these systems if the right ignition key is not used. Active anti-theft systems usually refer to a system that arms when the system is activated or when a sequence of events happens that automatically arm the system. An active system will sound the vehicle's horn(s) and disable the starter if it detects an attempt to break into one of its coverage zones. These zones can include the doors, trunk, hood, ignition, and a radio input. Some systems have active ultrasonic sound waves that set up an invisible shield to protect the interiors of convertibles.

Most common false alarms are caused by misplaced sensors or sensors that have been adjusted to an overly sensitive level, such as shock sensors. Many new shock and glass sensors now have two stage mechanisms: The sensor will give a warning when the first threshold is broken and will sound the alarm when the second threshold is broken. Door sensors will start to set false signals to the alarm module if they become rusted out or moving parts begin to wear out.

Alarm systems often tie into the interior dome light circuit to signal when a door has been opened. Some alarm systems are tied directly into the door ajar switch or have a switch in the door for specifically detecting when the door is open. The reason for a separate switch is that some manufacturers have door handle switches, instead of door switches, that signal a control module to illuminate the interior lights.

The remote keyless entry module is connected to the power door lock circuit. A small remote transmitter sends lock and unlock signals to this module when the appropriate buttons are pressed on the remote transmitter. When the handheld remote transmitter is a short distance from the vehicle, the module responds to the transmitter signals. When the unlock button is pressed on the remote transmitter, the module supplies voltage to the unlock relay winding to close these relay contacts and move the door lock motors to the unlock position. The batteries in the remote transmitters have to be replaced periodically.

When the unlock button is depressed on the remote transmitter, the locks will unlock and the interior lights will illuminate on most systems. Then the remote keyless entry module will turn off the interior lights after approximately one minute or when the ignition is turned on.

Some remote keyless entry systems have an option that will command the engine to start called "remote start". This system operates using the same technology as the system that unlocks the doors and operates the theft deterrent system. When the driver activates the remote start system, the body control module (BCM) sends a signal to the necessary devices to make the engine start and idle. The doors will remain locked until the driver approaches the vehicle and activates the unlock function. This system must be programmed in the same manner as the remote keyless entry system.

6. Inspect, test, and repair or replace components, connectors and wiring of security/anti-theft systems.

An anti-theft system is normally concealed within the vehicle, therefore it is difficult to visually inspect most of the components. Two important tools will allow the technician to properly test the system: a wiring diagram and a parts locator guide. The technician should treat the circuit just like any other electrical circuit and look for the type of problem before proceeding. A short will cause the fuse of the circuit breaker to blow. An open will prevent operation of the circuit. High resistance will cause the system to work improperly. All systems are unique and have a variety of sensors and controllers. Factory scan tools are essential when diagnosing theft system problems. These tools allow the technician to view live data and trouble codes in the anti-theft system.

7. Diagnose operation of entertainment and related circuits (such as: radio, DVD, remote CD changer, navigation, amplifiers, speakers, antennas, and voice activated accessories); determine needed repairs.

Audio systems typically work well, noisily, or not at all. Since these units are typically replaced rather than repaired, a technician simply identifies the faulty part and replaces it. If the unit does not work at all, the problem is most likely a lack of power to the unit or a poor ground.

Sound quality depends on a number of things. Rattles and buzzes are caused more often by loose speakers, speaker mountings, speaker grilles, or trim panels than by inoperative speakers. Check the tightness of all mounting and trim pieces when this type of noise is heard.

Sound distortion can be caused by the speaker, radio chassis, or wiring. If the concern is the chassis, all speakers on the same side of the vehicle will exhibit the same poor quality. Distortion can also be caused by damaged wiring, which is normally accompanied by lower than normal sound output.

Static may be caused by the charging system or the ignition system. A poor engine ground or poor ground at the sound system components may cause static in the sound. Defective radio suppression devices, such as a suppression coil on an instrument voltage limiter or a clamping diode on an electromagnetic clutch, may cause static on the radio. A defective antenna with poor ground shielding can also result in static.

Some vehicles are equipped with DVD entertainment systems that are built into the interior of the vehicle. The screens for these DVD systems can be found in the roof panel or in the seat back. The DVD player can be located near the radio controls in the front of the vehicle or located near the screen area in the rear passenger area. These systems also feature wireless headphones for the rear passengers to use to hear the sound from the unit. The batteries should be checked if the headphones do not perform as designated.

Voice activation systems are becoming more common on late-model vehicles. This system allows the driver to perform many tasks by using voice commands rather than pushing a button. The technology used in these systems is adaptive to speaking styles and patterns. Systems that can be manipulated with voice commands include the audio system, navigation system, climate control system and mobile phone system. Voice activation systems are integrated into the existing body electronic functions, which require use of a scan tool during troubleshooting.

8. Inspect, test, repair and/or replace components, connectors, and wiring of entertainment systems.

Repairing entertainment systems is similar to repairing other advanced electronic systems of the vehicle. Professional repair techniques should be used when repairing the wiring and connectors of the entertainment systems.

Some late-model vehicles are equipped with satellite radio systems. These fee-based radio systems receive signals from satellites through a special antenna. The reception on these systems can be interrupted when travelling near tall buildings and tree covered areas. The antenna for these entertainment systems is different from the standard radio found in the vehicles. If poor reception is experienced with one of these units, the technician should check the antenna for problems and also check the antenna lead wire.

9. Diagnose operation of safety systems and related circuits (such as: airbags, seat belt pretensioners, occupancy classification, wipers, washers, collision avoidance, passive speed control, heads-up display, park assist, and back up camera); determine needed repairs.

In many airbag systems, the warning light is illuminated for five to six seconds after the engine is started while the module performs system checks. The airbag warning light will then turn off if the system passes all its system tests. If the warning light stays illuminated, then there is a fault with the airbag system. When the airbag module is set to diagnostic mode, a code will be flashed through the airbag warning light. If this light should also fail, then the system may be set up to sound a warning tone through the warning chime system. The technician should also be able to retrieve trouble codes for the system with a diagnostic tool.

Many airbag systems are intuitive as to the number and size of the occupants in the front seats. The passenger seat is equipped with a pressure sensor that detects the weight of the occupant sitting in this location. If the occupant is below an average adult weight, then the airbag in this location will not deploy at the same level, or possibly not even deploy at all, during a frontal collision. In addition, the airbag will not deploy if there is not a person sitting in the passenger seat during a frontal collision. The sensor that is used on the occupant detection system can be diagnosed with a scan tool. Some of these sensors require a calibration that can be accomplished with a scan tool.

Some wiper motors contain a series field coil, a shunt field coil, and a relay. When the wiper switch is turned on, the relay winding is grounded through one set of switch contacts. This action closes the relay contacts. Current is supplied through these contacts to the series field coil and armature. Under this condition, the wiper motor starts turning. If the wiper switch is in the high-speed position, the shunt coil is not grounded and the motor turns at high speed.

When the wiper switch is in the low-speed position, the shunt coil is grounded through the second set of wiper switch contacts. Under this condition, current flows through the shunt coil and the wiper switch to ground. Current flow through the shunt coil creates a strong magnetic field that induces more opposing voltage in the armature windings. This opposing voltage in the armature windings reduces current flow through the series coil and armature windings to slow the armature. If the wiper motor fails to park or parks in the wrong position, the parking switch or cam is probably defective.

Some wiper motors have permanent magnets in place of the field coils. These motors have a low-speed and a high-speed brush. In some of these motors, the low-speed brush is directly opposite the common brush and the high-speed brush is positioned between these two brushes.

The most likely causes for constant operation of a wiper system are a faulty or stuck switch or the controlling circuit is shorted and the switch is by-passed. If the wipers do not park properly, the park mechanism inside the wiper motor may be at fault, the park circuit damaged, or the intermittent wiper unit damaged. If the wipers work intermittently, the most likely cause is a loose wire or connector. If the wipers do not work at all, the wiper motor is bad or there is an open in the circuit.

Many windshield washer systems have an electric pump mounted in the bottom of the washer fluid reservoir. When the washer button is pressed, voltage is supplied through the switch to the washer motor. This motor operates a pump that forces washer fluid through the hoses to the washer nozzles. Many late-model vehicles use the body control module (BCM) to control the washers. They use a sequence when the washer is activated that starts the wipers and sprays the windshield for a certain number of cycles. In older cars that ran sequences, this was accomplished with cams and the washer unit was physically attached to the wiper motor.

When checking the windshield washer system, check for low fluid levels and disconnected wires. Then try to isolate the problem by disconnecting the hose at the pump and operating the system. If the pump ejects a stream of fluid, then the hoses are clogged. If the pump does not spray, observe the pump motor while activating the washer switch. If the motor operates, check for blockage at the pump. If no blockage exists, replace the motor. If the motor fails to operate, check for voltage and ground at the motor. This will isolate the problem's location to the motor or the washer switch and wires.

Cruise control systems are often a combination of vacuum-operated devices, mechanical linkages, and electrical components. Always check for mechanical binding of the linkage before testing the vacuum and electrical components. Often there is a specified adjustment for the linkage. Check and adjust the linkage before doing more diagnostics. The vacuum devices can be checked for operation and leaks with a handheld vacuum pump. When vacuum is applied, a diaphragm should move. The unit should be able to hold any applied vacuum for quite some time. After the vacuum is released, the diaphragm should relax and return to its "off" position.

In some systems, the control module and the stepper motor are combined in one unit. A cable is connected from the stepper motor to the throttle linkage. The control unit receives inputs from the cruise control switch, brake switch, and vehicle speed sensor (VSS). The control module sends output commands to the stepper motor to provide the desired throttle opening. A defective VSS might cause erratic or no cruise control operation.

Many newer vehicles actuate an electric servo motor with a cable attached to the throttle body or linkage to control speed. The speed is controlled by the power train control module (PCM). Inputs such as vehicle speed sensor (VSS), brake switch, steering wheel inputs, cruise control set and on/off switches all feed information to the PCM which responds by opening the throttle and monitoring the speed through the VSS. Keep in mind that most late-model systems use the clock spring or spiral cable to transfer the commands from steering wheel mounted switches. Many times a vehicle with an inoperative cruise control will also show the air bag light on, if damage to the clock spring has occurred.

An option that has become common in many luxury vehicles is cruise control that is capable of slowing the vehicle down if a slower moving vehicle is approached. This system operates by using radar technology that projects radar signals in front of the vehicle. This system detects objects in the path of the vehicle and is capable of slowing the vehicle down to reduce the stress on the driver of the vehicle. Technicians should make sure that the transmitter and receiver are both clean and clear of obstructions.

Heads-up display systems project data from the instrument panel onto the windshield. The image that is projected appears as if it were some distance in front of the windshield. This system typically has a brightness control that the driver can adjust. It is built into the instrument panel and uses mirror technology to create the projected image. Technicians should use the normal diagnostic procedures on the instrument panel when diagnosing the heads-up display function.

Some manufacturers have systems on the vehicle that assist the driver in parking maneuvers in areas with limited room to back up. Such a system typically uses sensors that are located in the rear bumper to detect the proximity of objects. When the sensors detect a close object, the control module illuminates one or more indicators inside the vehicle to warn the driver. Some of the systems include an audible alarm that also warns the driver of the close proximity of an object. These park assist systems are very complex and have diagnostics integrated into their software. Technicians will need to use scan tools to retrieve trouble codes and view data from the proximity sensors. The proximity sensors should be kept clean from excess road dust and debris.

Back-up cameras are used on some vehicles to assist the driver when operating the vehicle in reverse. These systems typically include a camera that is mounted in the bumper and a screen that the driver uses to see what is behind the vehicle. Faults that can occur with the back-up camera system may be located in the camera, the display or the wiring harness. Technicians repairing problems on the back-up camera should use the same technique that is used on any other electronically controlled system. Inspect the camera and the display screen for physical damage. The wiring and connections should also be inspected for tightness and physical damage. A scan tool can be used to monitor the communication system for problems.

10. Inspect, test, and repair or replace components, connectors, and wiring of safety systems.

When servicing the airbag system, always disconnect the battery negative terminal first and wait for the manufacturer's specified time period to elapse for the system to power down entirely. This time period is usually one or two minutes. Never use a powered test light or an ohmmeter to diagnose an airbag system. Diagnose these systems with a voltmeter or the manufacturer's recommended diagnostic tool(s). Use of an ohmmeter should be restricted to circuits without connections to pyrotechnic devices. Since deployed air bags may contain residual chemicals, wear safety glasses and gloves when handling these components.

Sensors always should be mounted in their original orientation. Most sensors have directional arrows that must face in the specified direction. Front airbag sensors are positioned toward the front of the vehicle and side airbag sensors are aimed toward the sides.

When carrying inflator devices, be sure to hold them at a distance from the body and walk carefully to avoid falling. Always store inflator modules face up on the bench and carry these components with the trim cover facing away from the body.

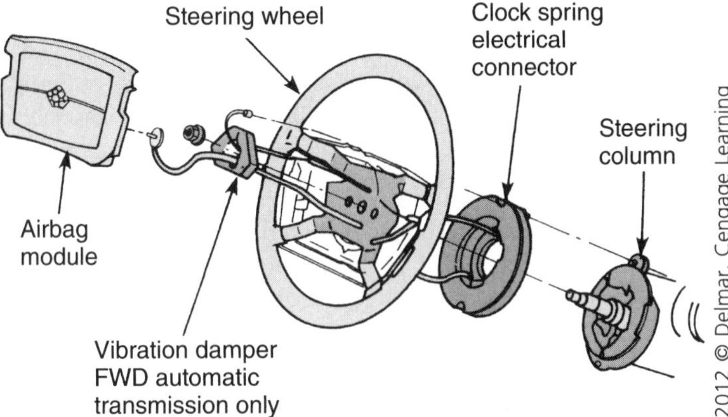

The exact procedure for disarming an airbag system varies from vehicle to vehicle; always refer to the appropriate service manual to do this operation. Typically, the procedure involves disconnecting the negative battery cable and taping the cable terminal to prevent accidental connection to the battery post. Then the supplemental inflatable restraint (SIR) fuse is removed from the fuse box. After this, the technician should wait at least 10 minutes to allow the reserve energy to dissipate before working on or around the air bag system.

SECTION

5 Sample Preparation Exams

INTRODUCTION

Included in this section are a series of six individual preparation exams that you can use to help determine your overall readiness to successfully pass the Electrical/Electronic Systems (A6) ASE certification exam. Located in Section 7 of this book you will find blank answer sheet forms you can use to designate your answers to each of the preparation exams. Using these blank forms will allow you to attempt each of the six individual exams multiple times without risk of viewing your prior responses.

Upon completion of each preparation exam, you can determine your exam score using the answer keys and explanations located in Section 6 of this book. Included in the explanation for each question is the specific task area being assessed by that individual question. This additional reference information may prove useful if you need to refer back to the task list located in Section 4 for additional support.

PREPARATION EXAM 1

1. A control module needs to be programmed following the replacement of the hardware. Technician A says that a multimeter will be needed to measure the resistance of the new software. Technician B says that a module programming database is needed to retrieve the correct program to upload to the new module. Who is correct?

   A. A only
   B. B only
   C. Both A and B
   D. Neither A nor B

2. Technician A says that voltage drop testing a connector needs to be done when the circuit is energized. Technician B says that voltage drop testing a relay needs to be done when the circuit is de-energized. Who is correct?

   A. A only
   B. B only
   C. Both A and B
   D. Neither A nor B

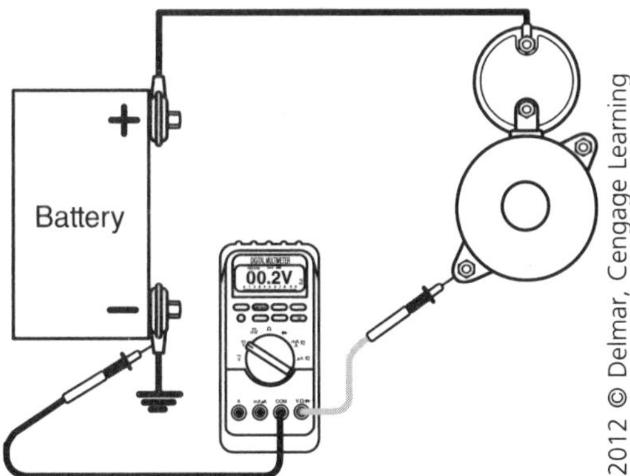

3. Referring to the figure, Technician A says that the test should be performed while cranking the engine. Technician B says that the reading on the meter indicates that the negative battery cable is faulty. Who is correct?
   A. A only
   B. B only
   C. Both A and B
   D. Neither A nor B

4. All of the following conditions can cause reduced current flow in an electrical circuit EXCEPT:
   A. Loose terminal connections
   B. A hot-side wire that is rubbing a metal component
   C. Corrosion inside the wire insulation
   D. Wire that is too small

5. Technician A says that an electrical switch that has continuity will have reduced current flow. Technician B says that a piece of wire that has high resistance will have increased current flow. Who is correct?
   A. A only
   B. B only
   C. Both A and B
   D. Neither A nor B

6. Technician A says that intermittent electrical signals can sometimes be diagnosed by using an oscilloscope. Technician B says that a broken tooth on a speed sensor reluctor can be sometimes be diagnosed by using an ohmmeter. Who is correct?
   A. A only
   B. B only
   C. Both A and B
   D. Neither A nor B

7. A vehicle is in the repair shop with an inoperative horn. A scan tool is connected and the horn operates when a functional test of the horn is performed. Technician A says that a faulty horn switch could be the problem. Technician B says that a faulty clock spring could be the problem. Who is correct?
   A. A only
   B. B only
   C. Both A and B
   D. Neither A nor B

8. A technician is replacing the electronic module for the park assist system. Which of the following practices would most likely be required to assure a trouble free repair?

    A. Remove the engine noise dampening cover.
    B. Disconnect the negative battery cable prior to beginning the repair.
    C. Remove the bumper retaining fasteners.
    D. Remove the rear hatch lock assembly.

9. A vehicle is in the repair shop with an inoperative power window. During diagnosis, a blown power window fuse is located. Technician A says that a short to ground between the switch and the motor could be the cause. Technician B says that a tight power window motor could be the cause. Who is correct?

    A. A only
    B. B only
    C. Both A and B
    D. Neither A nor B

10. A vehicle is in the repair shop for a dead battery. During the diagnosis, an excessive parasitic draw is found. Technician A says that a diode problem in the generator could be the cause. Technician B says that an "open" glove box bulb could be the problem. Who is correct?

    A. A only
    B. B only
    C. Both A and B
    D. Neither A nor B

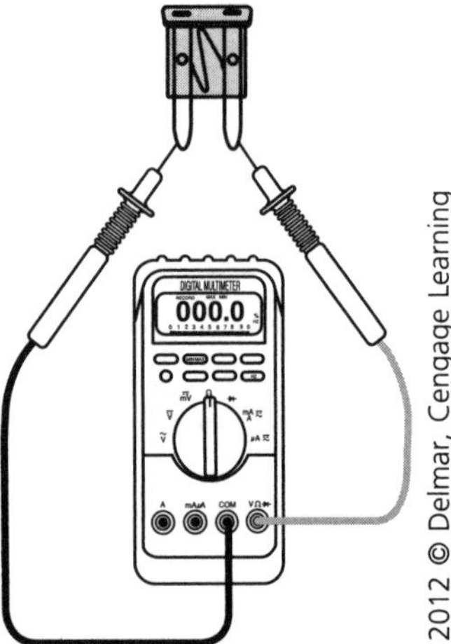

11. Referring to the figure above, Technician A says that the fuse is faulty because it has low resistance. Technician B says that the meter is set to test the resistance of the fuse. Who is correct?

    A. A only
    B. B only
    C. Both A and B
    D. Neither A nor B

12. Which of the following details would be the most likely item to be located on a wiring diagram?

    A. The location of a component
    B. A flowchart for troubleshooting an electrical problem
    C. The current rating of a circuit
    D. The color and circuit number of a wire

13. Technician A states that the scan tool receives data from a connector located on the ECM. Technician B states that the scan tool connects to the data bus using data link connector (DLC). Who is correct?

    A. A only
    B. B only
    C. Both A and B
    D. Neither A nor B

14. A vehicle is being diagnosed for a battery problem. Technician A says that the battery open circuit voltage should be 12.6 volts before performing a digital battery test. Technician B says that a battery can be accurately load-tested if it has an open circuit voltage of 12 volts. Who is correct?

    A. A only
    B. B only
    C. Both A and B
    D. Neither A nor B

15. Which of the following tools can be used to maintain electronic memory functions while a battery is disconnected?

    A. A battery jump box with a 12-volt accessory connecting wire
    B. A scan tool
    C. A digital multimeter
    D. A test light

16. All of the following details are critical when choosing a replacement battery for a vehicle EXCEPT:

    A. Cold cranking amps of the battery
    B. Reserve capacity of the battery
    C. Battery manufacturer
    D. Physical dimensions of the battery case

17. What is the most likely method to be used when professionally cleaning battery cable clamps?

    A. Pneumatic die grinder
    B. Wire brush
    C. Carburetor cleaner and scraper
    D. Screwdriver

18. A vehicle with a 4-cylinder engine is being diagnosed for a starting problem. A starter current draw test was performed and the amperage was 295 amps. Technician A says that worn starter brushes could be the problem. Technician B says that a shorted field coil could be the problem. Who is correct?

    A. A only
    B. B only
    C. Both A and B
    D. Neither A nor B

19. A vehicle is in the repair shop for a starting problem. There is a clicking sound at the starter when the key is moved to the start position. Technician A says that the starter solenoid could be the cause. Technician B says that worn starter brushes could be the problem. Who is correct?

    A. A only
    B. B only
    C. Both A and B
    D. Neither A nor B

20. Which of the following steps should the technician do first when beginning to remove a starter from a vehicle?

    A. Pull the starter fuse at the power distribution center.
    B. Remove the starter bolts.
    C. Disconnect the wires at the starter.
    D. Remove the negative battery cable and tape the terminal.

21. A vehicle is slow to crank. Technician A says that testing the voltage drop on the positive battery cable while cranking the engine will reveal potential problems in the battery. Technician B says that testing the voltage drop on the negative battery cable must be done with the ignition switch in the off position. Who is correct?

    A. A only
    B. B only
    C. Both A and B
    D. Neither A nor B

22. Technician A says that the starter solenoid pull-in winding can be tested by touching the "start" terminal and the "motor" terminal with the ohmmeter leads. Technician B says that the starter solenoid hold-in winding can be tested by touching the "start" terminal and the solenoid case with the ohmmeter leads. Who is correct?

    A. A only
    B. B only
    C. Both A and B
    D. Neither A nor B

23. All of the following conditions could cause an undercharged battery EXCEPT:
    A. Loose alternator bracket fasteners
    B. An oversized charging wire
    C. Worn generator brushes
    D. Excessive voltage drop in the charging wire

24. Technician A says that the charging system voltage will operate at approximately 11 to 13 volts. Technician B says that the charging system voltage should be measured while cranking the engine. Who is correct?
    A. A only
    B. B only
    C. Both A and B
    D. Neither A nor B

25. A vehicle is being diagnosed for a charging problem. The generator produced 85 amps during the output test and it is rated at 135 amps. Technician A says that the generator could have a bad diode. Technician B says that the generator drive pulley could be too small. Who is correct?
    A. A only
    B. B only
    C. Both A and B
    D. Neither A nor B

26. An inoperative generator is being diagnosed. After full-fielding the generator, it begins to charge at full capacity. Technician A says that the generator mounting bracket could be loose. Technician B says that a faulty voltage regulator could be the cause. Who is correct?
    A. A only
    B. B only
    C. Both A and B
    D. Neither A nor B

27. What is the most likely test that could prove that a poor connection was present at the charging insulated circuit?
    A. Voltage drop test of the charging ground circuit
    B. Voltage drop test of the negative battery cable
    C. Voltage drop test of the positive battery cable
    D. Voltage drop test of the charging output wire

28. A vehicle has a problem of the headlights on the right side not being as bright as the headlights on the left side. Technician A says that a bad dimmer switch could be the cause. Technician B says that a loose ground connection on the right side could be the cause. Who is correct?
    A. A only
    B. B only
    C. Both A and B
    D. Neither A nor B

29. Technician A says that the daytime running lights utilize the turn signal filaments on most late model vehicles. Technician B says that the daytime running lights will use more current than the headlights. Who is correct?

    A. A only
    B. B only
    C. Both A and B
    D. Neither A nor B

30. The fog lights are dimmer than normal and a voltage drop test across the bulbs reveals 8 volts. Which of the following conditions would most likely cause this problem?

    A. Faulty fog light bulbs
    B. Faulty fog light switch contacts
    C. Faulty fog light fuse
    D. Faulty turn signal switch

31. All of the following conditions could cause the stop lights to be inoperative EXCEPT:

    A. A closed stop light switch
    B. Open stop light switch
    C. Blown stop light fuse
    D. Open turn signal switch

32. Technician A says that turn signal flashers are sensitive to electrical load. Technician B says that hazard flashers are not sensitive to electrical load. Who is correct?

    A. A only
    B. B only
    C. Both A and B
    D. Neither A nor B

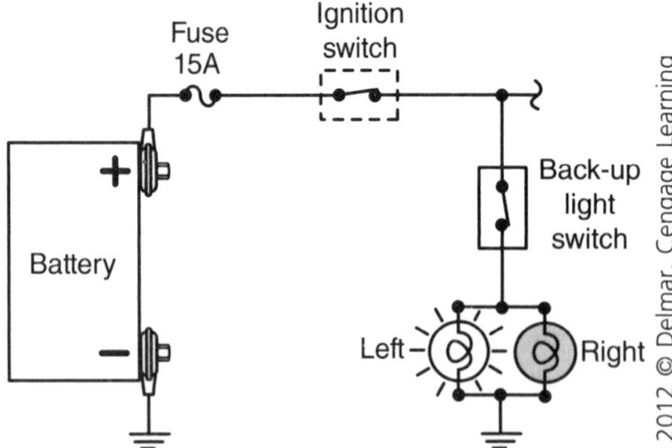

33. Referring to the figure, the back-up lamp on the right side is inoperative but the left side works fine. Technician A says that the right side lamp socket could be open. Technician B says that the right side bulb could be open. Who is correct?

    A. A only
    B. B only
    C. Both A and B
    D. Neither A nor B

34. A vehicle is being diagnosed that has a fuel gauge that does not work. Technician A says that the other gauges should be checked for correct operation as part of the diagnosis process. Technician B says that the sending unit for the fuel gauge is in the fuel tank. Who is correct?

A. A only
B. B only
C. Both A and B
D. Neither A nor B

35. What type of sending unit does the temperature gauge use?

A. Rheostat
B. Potentiometer
C. Photo resistor
D. Thermistor

36. A vehicle is being diagnosed for a problem of all of the gauges being inoperative. The vehicle is equipped with an electronic cluster that has self-diagnostic capabilities. Technician A says that a scan tool should be connected to the data link connector to retrieve cluster data. Technician B says that this problem could be diagnosed by using the self-diagnostic function of the instrument cluster. Who is correct?

A. A only
B. B only
C. Both A and B
D. Neither A nor B

37. All of the following types of electronic devices are used as inputs to electronic gauge assemblies EXCEPT:

A. Thermistor
B. Piezo resistor
C. Diode
D. Rheostat

38. The temperature light does not illuminate during the bulb test or when the vehicle overheats. The most likely cause for this condition would be?

A. Temperature sending unit
B. Bulb is open
C. Grounded wire to the sending unit
D. Open wire to the sending unit

39. All of the following items affect the driver information center EXCEPT:

A. Oxygen sensor
B. Power train control module
C. Body control module
D. Data bus network

40. Technician A says that the horn relay can be activated by the BCM. Technician B says that the horn relay can be activated by a ground signal that is received from the horn switch. Who is correct?

A. A only
B. B only
C. Both A and B
D. Neither A nor B

41. The horn does not operate when the horn switch is depressed but it will operate when the panic button is depressed on the keyless entry transmitter. Technician A says that the relay coil could be shorted. Technician B says that the horn fuse could be blown. Who is correct?

A. A only
B. B only
C. Both A and B
D. Neither A nor B

42. What is the most likely problem that could cause the windshield wipers to continue running after the wiper switch is turned off?

A. Open wire between the wiper switch and the wiper motor
B. Shorted wiper wash motor switch
C. Shorted wiper park switch
D. Open wiper relay

43. All of the power windows operate slowly in both the up and down positions. Technician A says that the ground for the master switch could have a loose connection. Technician B says that a wrong fuse could be installed for the circuit. Who is correct?

A. A only
B. B only
C. Both A and B
D. Neither A nor B

44. A "scissor style" power window regulator is being replaced on a vehicle. Technician A says this can be completed without removing the motor from the door skin. Technician B says care should be taken when disconnecting the motor from the regulator due to the spring tension on the regulator. Who is correct?

A. A only
B. B only
C. Both A and B
D. Neither A nor B

45. Technician A says that the rear window defogger uses heated air to defog the back glass. Technician B says that the rear window defogger operates on a timer in order to limit the heavy electrical load on the vehicle. Who is correct?

A. A only
B. B only
C. Both A and B
D. Neither A nor B

46. The keyless entry system does not work with remote transmitter #1 but works correctly with remote transmitter #2. Technician A says that the battery in transmitter #1 may be weak and need to be replaced. Technician B says that transmitter #1 could be defective and need to be replaced. Who is correct?

 A. A only
 B. B only
 C. Both A and B
 D. Neither A nor B

47. The radio has poor reception and has a buzzing sound that increases with engine speed. The operation of the sound system is normal when a CD is used in the player. Technician A says that the main sound system ground could have a loose connection. Technician B says that the antenna coaxial cable could be damaged. Who is correct?

 A. A only
 B. B only
 C. Both A and B
 D. Neither A nor B

48. The auxiliary power outlet is inoperative and the fuse is found to be open. What is the most likely cause for this condition?

 A. Loose connection at the power outlet plug
 B. Foreign metal object in the power outlet
 C. Broken wire leading to the power outlet
 D. Open internal connection at the power outlet

49. All of the following conditions could cause the airbag light to stay on while the vehicle is in operation EXCEPT:

 A. An open clock spring
 B. A shorted airbag inflator
 C. A faulty seatbelt buckle
 D. An open airbag light bulb

50. An airbag system needs to be disarmed during a dash pad replacement. Technician A says that it is necessary to remove power from the system prior to component removal. Technician B says that the inflatable devices should be laid "face up" in a secure area after being removed from a vehicle. Who is correct?

 A. A only
 B. B only
 C. Both A and B
 D. Neither A nor B

PREPARATION EXAM 2

1. A late-model luxury vehicle is being diagnosed for a problem in the driver's side heated seat. The seat stays hot at all times regardless of the seat heater switch position. Technician A says that a shorted seat heater relay could cause this problem. Technician B says that a grounded circuit between the seat heater relay and the body control module (BCM) could be the cause. Who is correct?

   A. A only
   B. B only
   C. Both A and B
   D. Neither A nor B

2. All of the following statements about performing voltage drop tests on electrical circuits are true EXCEPT:

   A. The voltmeter leads should be connected on each end of the circuit to be tested.
   B. The meter should be connected in series with the circuit.
   C. The circuit should be energized when performing a voltage drop test.
   D. The digital meter should be set to the DC volts scale.

3. Technician A says that a poor ground connection will cause reduced current flow in an electrical circuit. Technician B says that water corrosion in the wiring will cause reduced current flow in an electrical circuit. Who is correct?

   A. A only
   B. B only
   C. Both A and B
   D. Neither A nor B

4. A resistance test was performed on a closed brake switch and the result is .857 megohms. Technician A says that the switch is faulty because the reading is beyond the specifications. Technician B says that the switch has 857 ohms of resistance. Who is correct?

   A. A only
   B. B only
   C. Both A and B
   D. Neither A nor B

5. Technician A says that an oscilloscope shows voltage signals represented as lines on a digital screen. Technician B says that an oscilloscope can be used to view live sensor readings directly at each sensor. Who is correct?

   A. A only
   B. B only
   C. Both A and B
   D. Neither A nor B

6. Technician A says that a scan tool can be used to retrieve a trouble code from the body control module. Technician B says that a scan tool can be used to view live sensor data from the body control module. Who is correct?

   A. A only
   B. B only
   C. Both A and B
   D. Neither A nor B

7. Voice activation commands are used on some luxury vehicles to interact with all of the following accessory systems EXCEPT:

   A. Power windows
   B. Audio system
   C. Navigation system
   D. Mobile phone system

8. Technician A says that a broken wire will often cause a fuse to blow due to high current flow. Technician B says that a corroded wire connection will often cause a fuse to blow due to high current flow. Who is correct?

   A. A only
   B. B only
   C. Both A and B
   D. Neither A nor B

9. A vehicle is in the repair shop for a dead battery. During the diagnosis, an excessive parasitic draw is found. Technician A says that removing one fuse at a time is an effective method of locating the unwanted draw. Technician B says that a "stuck closed" trunk light switch could be the cause. Who is correct?

   A. A only
   B. B only
   C. Both A and B
   D. Neither A nor B

10. Technician A says that tugging on a fusible link is an acceptable method of quickly testing it. Technician B says that a fusible link can be tested with an ohmmeter without disconnecting it from the vehicle. Who is correct?

    A. A only
    B. B only
    C. Both A and B
    D. Neither A nor B

11. Which of the following conditions would LEAST LIKELY cause a fusible link for the power windows to burn up?

    A. A motor with a shorted winding
    B. A power-side short to ground
    C. High resistance in the switch contacts
    D. A binding window regulator

12. Which of the following details would be the LEAST LIKELY item to be located on a wiring diagram?

    A. The circuit number of a wire
    B. A flowchart for troubleshooting an electrical problem
    C. The amp rating of the fuse that supplies the circuit
    D. The color of a wire

13. Technician A states that the scan tool receives data from the data bus network. Technician B states that a bi-directional scan tool can send functional messages over the data bus network. Who is correct?

    A. A only
    B. B only
    C. Both A and B
    D. Neither A nor B

14. A vehicle is in the repair shop for a battery problem. Technician A says that the battery voltage should be 12.4 volts before performing a battery load test. Technician B says that a battery can be accurately tested with a digital tester if it has at least 12 volts. Who is correct?

    A. A only
    B. B only
    C. Both A and B
    D. Neither A nor B

15. A battery is being replaced in a vehicle. Technician A says that the replacement battery should have at least 75 percent of the original battery CCA rating for the vehicle. Technician B says that the replacement battery should not exceed the physical dimensions of the original battery. Who is correct?

    A. A only
    B. B only
    C. Both A and B
    D. Neither A nor B

16. Technician A says that batteries can be recharged more quickly by using a high setting on the battery charger. Technician B says that batteries can be more thoroughly charged by using a low setting on the battery charger. Who is correct?

    A. A only
    B. B only
    C. Both A and B
    D. Neither A nor B

17. A vehicle needs to be jumpstarted with an auxiliary power supply. Technician A says that safety glasses should always be worn when working around batteries. Technician B says that the positive connection should be made at the starter solenoid. Who is correct?

    A. A only
    B. B only
    C. Both A and B
    D. Neither A nor B

18. Technician A says that a shorted armature can cause elevated starter current draw. Technician B says that a corroded battery cable terminal can cause elevated starter current draw. Who is correct?

    A. A only
    B. B only
    C. Both A and B
    D. Neither A nor B

19. A vehicle with a six cylinder engine is being diagnosed for a starting problem. A starter current draw test was performed and the amperage was 310 amps. Technician A says that this reading could be caused by a weak battery. Technician B says that this reading could be caused by tight starter bushings. Who is correct?

    A. A only
    B. B only
    C. Both A and B
    D. Neither A nor B

20. All of the following components are parts of the starter control circuit EXCEPT:

    A. Park/neutral switch
    B. Positive battery cable
    C. Starter relay
    D. Ignition switch

21. Technician A says that the starter solenoid provides the linear movement to push the starter drive gear into the flywheel. Technician B says that the starter solenoid acts as a magnetic switch to provide a current path for the positive circuit to reach the starter motor. Who is correct?

    A. A only
    B. B only
    C. Both A and B
    D. Neither A nor B

22. Technician A says that the negative battery cable should be removed prior to disconnecting the electrical connections at the starter. Technician B says that the fasteners should be removed prior to removing the electrical connections at the starter. Who is correct?

    A. A only
    B. B only
    C. Both A and B
    D. Neither A nor B

23. Which of the following conditions would be the LEAST LIKELY cause of an undercharged battery?

    A. Worn generator brushes
    B. Loose fastener at the charge wire connection
    C. Faulty voltage regulator
    D. Battery cable that is oversized

24. The accessory drive belt system should be inspected during regular intervals. Technician A says that a serpentine drive belt tensioner should snap back after releasing pressure on it. Technician B says that the drive belt should be replaced at the first sign of cracks on the back side of it. Who is correct?
   A. A only
   B. B only
   C. Both A and B
   D. Neither A nor B

25. What is the LEAST LIKELY cause for low charging current during a charging output test?
   A. Worn brushes
   B. Faulty rectifier bridge
   C. Shorted stator winding
   D. Faulty charging gauge

26. A vehicle with a charging problem is being repaired. The charging output wire received damage and needed to be repaired. Technician A says to use weather resistant connectors when making wire repairs in the engine compartment. Technician B says that solder and heat shrink is an acceptable method of wire repair in the engine compartment. Who is correct?
   A. A only
   B. B only
   C. Both A and B
   D. Neither A nor B

27. A generator is being replaced. Technician A uses an air tool to remove the fastening nut from the charging output wire. Technician B disconnects the negative battery cable prior to removing the generator. Who is correct?
   A. A only
   B. B only
   C. Both A and B
   D. Neither A nor B

28. Technician A uses an alignment tool to aim the headlights on a late model vehicle. Technician B uses the built-in alignment bubbles on the headlamp housings to aim the headlights on some late model vehicles. Who is correct?
   A. A only
   B. B only
   C. Both A and B
   D. Neither A nor B

29. The left rear park light bulb is inoperative. The turn signal and stop lamp bulbs work normally on that side. Technician A says that a faulty ground on the left rear could be the cause. Technician B says that a faulty bulb could be the cause. Who is correct?
   A. A only
   B. B only
   C. Both A and B
   D. Neither A nor B

30. A vehicle electronic instrument cluster lighting intermittently goes dark after being driven for long periods. Technician A says the problem could be a shorted LED. Technician B says the problem could be a faulty lighting driver in the instrument cluster assembly. Who is correct?

A. A only

B. B only

C. Both A and B

D. Neither A nor B

31. A vehicle is being diagnosed for inoperative brake lights. Technician A says that a blown stop light fuse could be the problem. Technician B says that a shorted headlight switch could be the cause. Who is correct?

A. A only

B. B only

C. Both A and B

D. Neither A nor B

32. A vehicle is being diagnosed for inoperative turn signals. The hazard lights work correctly. Technician A says that a faulty turn signal flasher could be the cause. Technician B says that a faulty multi-function switch could be the cause. Who is correct?

A. A only

B. B only

C. Both A and B

D. Neither A nor B

33. Technician A says that the hazard flasher on late-model vehicles can be replaced without disconnecting the vehicle battery. Technician B says that the turn signal flasher and hazard flasher are combined into one assembly on some late-model vehicles. Who is correct?

A. A only

B. B only

C. Both A and B

D. Neither A nor B

34. The back-up light switch is being tested on a vehicle. Technician A says that the back-up light switch is often combined with the turn signal switch. Technician B says that the back-up light switch is located near the rear light housing. Who is correct?

A. A only

B. B only

C. Both A and B

D. Neither A nor B

35. A vehicle is being diagnosed with a problem of the door ajar alarm sounding intermittently. Technician A says that the striker adjustment of all doors should be checked. Technician B says that the door jam switches should be replaced. Who is correct?

A. A only

B. B only

C. Both A and B

D. Neither A nor B

36. What type of sending unit does the oil pressure gauge use?

    A. Rheostat
    B. Potentiometer
    C. Photo resistor
    D. Piezo resistor

37. Which of the following devices is LEAST LIKELY to be used as an input to an electronic gauge assembly?

    A. Thermistor
    B. Piezo resistor
    C. Body control module
    D. Light emitting diode (LED)

38. The temperature light does not illuminate during the bulb test or when the vehicle overheats. Technician A says that a faulty instrument cluster unit could be the cause. Technician B says that a blown bulb could be the cause. Who is correct?

    A. A only
    B. B only
    C. Both A and B
    D. Neither A nor B

39. Technician A says that the body control module (BCM) communicates with electronic driver information centers. Technician B says that the power train control module (PCM) communicates information to the electronic driver information center. Who is correct?

    A. A only
    B. B only
    C. Both A and B
    D. Neither A nor B

40. An electronic chime control module needs to be replaced on a late model vehicle. Technician A says that these modules are located in the engine compartment. Technician B says that the body control module (BCM) has to be reprogrammed after replacing the chime control module. Who is correct?

    A. A only
    B. B only
    C. Both A and B
    D. Neither A nor B

41. A vehicle has a problem with the delay windshield wiper operation. The wipers work on low speed and high speed but do not work at all when the delay function is chosen. What is the most likely cause of this problem?

    A. Washer motor
    B. Potentiometer in the delay switch
    C. Wiper motor low speed brush
    D. Wiper motor high speed brush

42. A vehicle is having the wiper motor replaced. All of the following steps will need to be performed EXCEPT:

    A. Remove the wiper arms.
    B. Remove the cowl panel.
    C. Remove the wiper motor fasteners.
    D. Remove the vehicle battery from the engine compartment.

43. The windshield washer pump motor runs continuously while the ignition switch is on. Technician A says that the multi-function switch could be shorted. Technician B says that the wiper control module could be defective. Who is correct?

    A. A only
    B. B only
    C. Both A and B
    D. Neither A nor B

44. A vehicle with heated seats has a problem on the driver's side. The heated seat only gets mildly warm when the switch is turned on. Technician A says that the driver's side heated seat relay could have burnt contacts. Technician B says the driver's side heated seat relay could have an open coil. Who is correct?

    A. A only
    B. B only
    C. Both A and B
    D. Neither A nor B

45. One section of a rear defogger does not clear the glass when in operation. The rest of the glass gets cleared normally when the rear defogger is used. Technician A says that the rear defogger relay could be defective. Technician B says that one of the heat strips could have a break in it. Who is correct?

    A. A only
    B. B only
    C. Both A and B
    D. Neither A nor B

46. The power door lock only works intermittently on the passenger front door. All of the other doors work normally. Technician A says that the wires could be damaged at the point near where the door connects to the vehicle body. Technician B says that the passenger door lock actuator could have a poor electrical connection. Who is correct?

    A. A only
    B. B only
    C. Both A and B
    D. Neither A nor B

47. The keyless entry control module is being replaced on a late-model vehicle. Technician A says that all of the transmitters will work with the new controller without having to be reprogrammed. Technician B says that the power door locks will have to be cycled multiple times to reprogram the keyless entry control module. Who is correct?

   A. A only
   B. B only
   C. Both A and B
   D. Neither A nor B

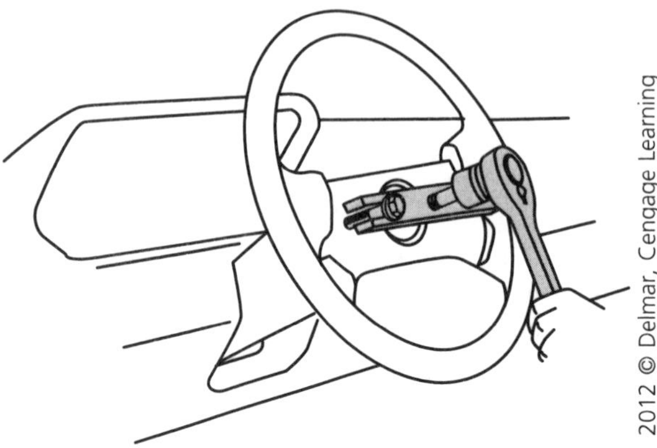

48. Which cruise control system component is most likely being serviced in the figure above?

   A. Clock spring/ribbon wire
   B. Servo
   C. Control module
   D. Cancel switch

49. Technician A says that audio speakers can be checked by testing the resistance with an ohmmeter. Technician B says that an antenna lead wire can be checked by testing the resistance with an ohmmeter. Who is correct?

   A. A only
   B. B only
   C. Both A and B
   D. Neither A nor B

50. A digital clock is being diagnosed for losing the correct time each time the ignition is shut off. Technician A says that the main ground connection for the clock could be disconnected. Technician B says that the "keep alive" fuse could be blown. Who is correct?

   A. A only
   B. B only
   C. Both A and B
   D. Neither A nor B

PREPARATION EXAM 3

1. Which of the following practices would be the LEAST LIKELY use of a 12 volt test light?
   - A. Checking for power at a rear defogger grid
   - B. Checking for power at a headlight bulb
   - C. Checking for reference voltage at temperature sending unit
   - D. Checking the input and output voltage of a fuse at the fuse panel

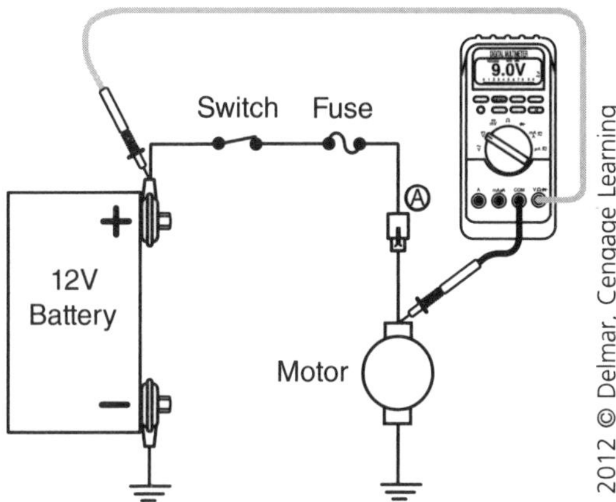

2. Referring to the 12 volt circuit above, the motor runs slowly when the switch is turned on. Technician A says that a faulty motor ground could cause this condition. Technician B says that faulty terminal at connector A could be the cause. Who is correct?
   - A. A only
   - B. B only
   - C. Both A and B
   - D. Neither A nor B

3. Technician A says that a poor ground connection will cause increased current flow in an electrical circuit which would likely blow a fuse. Technician B says that water corrosion in the wiring will cause reduced current flow in an electrical circuit. Who is correct?
   - A. A only
   - B. B only
   - C. Both A and B
   - D. Neither A nor B

4. Which of the following methods of testing amperage in a circuit would be LEAST LIKELY to interrupt the circuit?
   - A. Remove the switch and connect the ammeter leads to the exposed terminals
   - B. Connecting an amp clamp around the wire
   - C. Remove the relay and connect the ammeter leads across the control side of the circuit
   - D. Connecting the voltmeter across the fuse

5. A resistance test was performed on an open headlight switch and the result is 3.857 megohms. Technician A says that the switch is faulty because the reading is beyond the specifications. Technician B says that the switch has over 3 million ohms of resistance. Who is correct?

   A. A only
   B. B only
   C. Both A and B
   D. Neither A nor B

6. Technician A says that an oscilloscope can be used as a scan tool to read live data and codes. Technician B says that an oscilloscope can be used to view live sensor readings directly at each sensor. Who is correct?

   A. A only
   B. B only
   C. Both A and B
   D. Neither A nor B

7. Technician A says that a scan tool can be used to retrieve a trouble code from the engine control module. Technician B says that a scan tool can be used to view live sensor data from the engine control module. Who is correct?

   A. A only
   B. B only
   C. Both A and B
   D. Neither A nor B

8. Technician A says that a fused jumper wire can be used to provide power to the horn during diagnosis. Technician B says that a fused jumper wire can be used to bypass a rear defogger relay during the diagnosis of that circuit. Who is correct?

   A. A only
   B. B only
   C. Both A and B
   D. Neither A nor B

9. Technician A says that a wire that is shorted to ground before the load will cause a fuse to blow due to high current flow. Technician B says that a corroded wire connection will cause the electrical load to receive increased current flow. Who is correct?

   A. A only
   B. B only
   C. Both A and B
   D. Neither A nor B

10. All of the following procedures are acceptable methods of locating excessive parasitic draw EXCEPT:

   A. Disconnect the battery ground cable from the engine block.
   B. Remove the fuses one at a time while watching the ammeter.
   C. Inspect the whole vehicle for any lamp that could be staying on.
   D. Disconnect the charge wire at the generator while watching the ammeter.

11. Which of the following conditions would LEAST LIKELY cause a circuit breaker to open?

    A. A stalled motor
    B. A power-side short to ground
    C. A ground-side short to ground
    D. A motor with a shorted winding

12. All of the following items are typically found in a wiring diagram EXCEPT:

    A. The color of the wire
    B. The circuit number of a wire
    C. The connector numbers for the circuit in question
    D. The locations of each splice used in the circuit

13. Technician A says that a high impedance digital meter is needed to perform voltage tests on the data bus network. Technician B says that an oscilloscope can be used to view the communication activity on the data bus network. Who is correct?

    A. A only
    B. B only
    C. Both A and B
    D. Neither A nor B

14. Technician A says that the battery voltage should be 12.4 volts before performing a load test. Technician B says that a battery state of charge can be determined by measuring the open circuit voltage across the terminals. Who is correct?

    A. A only
    B. B only
    C. Both A and B
    D. Neither A nor B

15. A battery is being replaced in a vehicle. Technician A says that the replacement battery should have at least the minimum CCA rating for the vehicle. Technician B says that the replacement battery should not exceed the reserve capacity of the original battery. Who is correct?

    A. A only
    B. B only
    C. Both A and B
    D. Neither A nor B

16. Technician A says that batteries can be recharged more quickly by using a low setting on the battery charger. Technician B says that batteries can be more thoroughly charged by using a high setting on the battery charger. Who is correct?

    A. A only
    B. B only
    C. Both A and B
    D. Neither A nor B

17. A vehicle needs to be jumpstarted with an auxiliary power supply. Technician A says that safe practices should always be followed when working around batteries. Technician B says that the negative connection should be made at a secure engine ground. Who is correct?

    A. A only
    B. B only
    C. Both A and B
    D. Neither A nor B

18. Technician A says that burnt starter solenoid contacts can cause elevated starter current draw. Technician B says that a partially cut battery cable can cause elevated starter current draw. Who is correct?

    A. A only
    B. B only
    C. Both A and B
    D. Neither A nor B

19. Which of the following components is LEAST LIKELY to be part of the starter control circuit.

    A. Positive battery cable
    B. Ignition switch
    C. Park/neutral switch
    D. Starter relay

20. Technician A says that the starter solenoid contacts can be tested by performing a voltage drop test across the "bat" and "motor" terminals when the starter is disengaged. Technician B says that the pull-in winding can be tested by measuring the resistance from the "S" terminal to the "motor" terminal. Who is correct?

    A. A only
    B. B only
    C. Both A and B
    D. Neither A nor B

21. Technician A says that the airbag fuse should be removed prior to disconnecting the electrical connections at the starter. Technician B says that the starter can be supported by the electrical wires without any expected damage to the wires. Who is correct?

    A. A only
    B. B only
    C. Both A and B
    D. Neither A nor B

22. Technician A says that a mistimed engine can cause slow cranking speed. Technician B says that low engine compression can cause rapid cranking speed. Who is correct?

    A. A only
    B. B only
    C. Both A and B
    D. Neither A nor B

23. Technician A says that a poor connection at the charging output wire can cause a charging problem. Technician B says that many charging circuits contain a circuit protection device. Who is correct?

    A. A only
    B. B only
    C. Both A and B
    D. Neither A nor B

24. The accessory drive belt system should be inspected during regular intervals. Technician A says that a serpentine drive belt tensioner should maintain spring pressure on the belt at all times. Technician B says that the drive belt should be replaced if major pieces are missing from the drive surface. Who is correct?

    A. A only
    B. B only
    C. Both A and B
    D. Neither A nor B

25. A vehicle is being diagnosed for a charging problem. The generator produced 50 amps during the output test and it is rated at 105 amps. Technician A says that the generator could have a worn brushes. Technician B says that the generator drive belt could be slipping. Who is correct?

    A. A only
    B. B only
    C. Both A and B
    D. Neither A nor B

26. All of the following methods of wire repair in the charging system are currently used EXCEPT:

    A. Crimp and seal connectors
    B. Butt connectors and tape
    C. New wiring harness
    D. Solder and heat shrink

27. A generator is being replaced. Technician A uses an air tool to install the fastening nut onto the charging output wire. Technician B carefully routes the drive belt around all of the pulleys before releasing the belt tensioner. Who is correct?

    A. A only
    B. B only
    C. Both A and B
    D. Neither A nor B

28. The dimmer switch has failed and needs to be replaced on a late-model vehicle. Technician A says that some dimmer switches are built into the multi-function switch. Technician B says that the replacing some dimmer switches requires the removal of the steering wheel. Who is correct?

    A. A only
    B. B only
    C. Both A and B
    D. Neither A nor B

29. Which of the following devices regulates the brightness of the dash lights?

A. Positive temperature coefficient
B. Door jam switch
C. Dimmer switch
D. Rheostat

30. The courtesy lights stay on continuously and have caused the battery to be discharged while the vehicle is parked for long periods of time. Technician A says that the courtesy light switch could have a bad connection. Technician B says that a door ajar switch could be shorted. Who is correct?

A. A only
B. B only
C. Both A and B
D. Neither A nor B

31. All of the following conditions could cause the brake lights to be inoperative EXCEPT:

A. Brake switch is stuck "open."
B. Brake light fuse is blown.
C. Brake light bulbs are blown.
D. Battery cable is loose.

32. A vehicle is being diagnosed for inoperative turn signals. All of the hazard lights work correctly. Technician A says that a blown turn signal bulb could be the cause. Technician B says that an open ground connection at the right rear lamp socket could be the cause. Who is correct?

A. A only
B. B only
C. Both A and B
D. Neither A nor B

33. A vehicle needs to have a trailer wiring harness installed into the existing lighting harness. Technician A says that many trailer wiring companies provide harnesses that tee directly into the existing harness. Technician B says that the turn signal flasher will need to be replaced with a heavy duty flasher. Who is correct?

A. A only
B. B only
C. Both A and B
D. Neither A nor B

34. A vehicle is being diagnosed with a problem of the door ajar alarm sounding intermittently. The vehicle is driven for several miles without duplicating the problem. Technician A says that the striker adjustment of all doors should be checked. Technician B says that the door jam switches adjustment should be checked. Who is correct?

A. A only
B. B only
C. Both A and B
D. Neither A nor B

35. A vehicle is being diagnosed with a fuel gauge that does not work. Technician A says that the fuel tank should be removed to test the fuel sending unit. Technician B says that the instrument cluster should be removed to inspect the connections. Who is correct?

    A. A only
    B. B only
    C. Both A and B
    D. Neither A nor B

36. What type of sending unit does the fuel gauge use?

    A. Rheostat
    B. Transducer
    C. Photo resistor
    D. Thermistor

37. A vehicle is being diagnosed for a problem of all of the gauges being inoperative. The vehicle is equipped with an electronic cluster that has self-diagnostic capabilities. Technician A says that every sending unit should be checked individually. Technician B says that this problem could be diagnosed by using the self-diagnostic function of the instrument cluster. Who is correct?

    A. A only
    B. B only
    C. Both A and B
    D. Neither A nor B

38. The temperature light stays on continuously while driving the vehicle. Technician A says that a shorted wire leading to the temperature sending unit could be the cause. Technician B says that an "open" temperature sending unit could be the cause. Who is correct?

    A. A only
    B. B only
    C. Both A and B
    D. Neither A nor B

39. Technician A says that the electronic driver information center receives data from the vehicle data bus network. Technician B says that the electronic driver information center can display "data bus" messages to the driver or technician. Who is correct?

    A. A only
    B. B only
    C. Both A and B
    D. Neither A nor B

40. A vehicle has a problem with the delay windshield wiper operation. The wipers work on low speed and high speed but do not work at all when the delay function is chosen. Technician A says that the main ground connection for the wiper system could be poorly connected. Technician B says that the wiper control module could be faulty. Who is correct?

    A. A only
    B. B only
    C. Both A and B
    D. Neither A nor B

41. A vehicle is having the wiper motor replaced. Which of the following steps is LEAST LIKELY to be performed?

    A. Remove the wiper arms.
    B. Remove the accessory drive belt.
    C. Remove the wiper motor fasteners.
    D. Remove the cowl panel.

42. The windshield washer pump is totally inoperative. Technician A says that the wiper switch could be defective. Technician B says that the wiper control module could be defective. Who is correct?

    A. A only
    B. B only
    C. Both A and B
    D. Neither A nor B

43. A vehicle with heated seats has a problem on the passenger side. The heated seat only gets mildly warm when the switch is turned on. Technician A says that the passenger side heated seat relay could have open contacts when turned on. Technician B says the passenger side heated seat heat strip could have a poor connection. Who is correct?

    A. A only
    B. B only
    C. Both A and B
    D. Neither A nor B

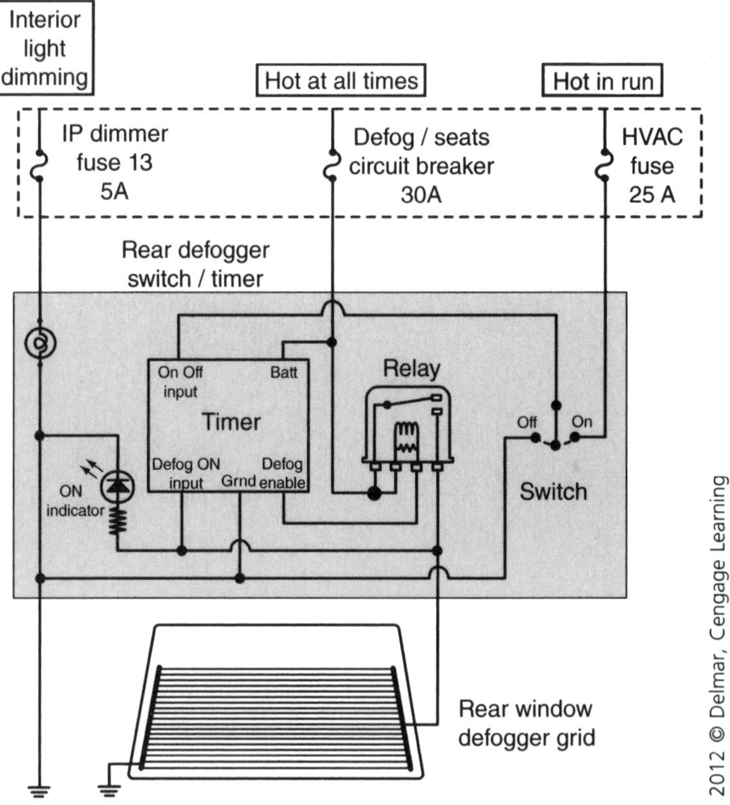

44. Referring to the figure above, the rear defogger does not clear the back glass when the system is turned on. The "on indicator" illuminates when the system is turned on. Technician A says that an open ground at the rear defogger grid could be the cause. Technician B says that an open grid line could be the cause. Who is correct?

 A. A only
 B. B only
 C. Both A and B
 D. Neither A nor B

45. The power door lock at the driver's door does not work at all. All of the other actuators work normally. Technician A says that the wires could be damaged at the point near where the door connects to the vehicle body. Technician B says that the driver's door lock actuator could be faulty. Who is correct?

 A. A only
 B. B only
 C. Both A and B
 D. Neither A nor B

46. All of the following conditions could cause the sunroof to operate slowly EXCEPT:

 A. Burnt relay contacts
 B. Tight sunroof motor linkage
 C. Binding sunroof tracks
 D. Using a 40 amp fuse in place of a 30 amp fuse

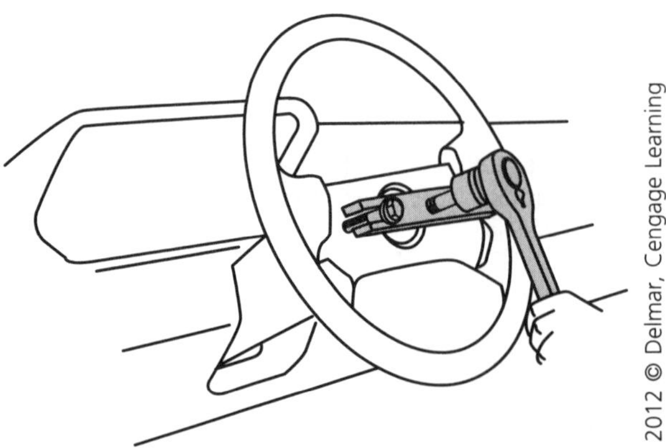

2012 © Delmar, Cengage Learning

47. Which airbag system component is most likely being serviced in the figure above?

    A. Crash sensor
    B. Clock spring
    C. Control module
    D. Arming sensor

48. Technician A says that some power antenna systems utilize a relay to raise the power antenna when the radio is turned on. Technician B says that some power antenna systems utilize a relay to lower the power antenna when the radio is turned off. Who is correct?

    A. A only
    B. B only
    C. Both A and B
    D. Neither A nor B

49. The cruise control is inoperative on a late-model vehicle. Technician says that a broken speedometer cable is the likely cause. Technician B says that a misadjusted throttle plate could be the cause. Who is correct?

    A. A only
    B. B only
    C. Both A and B
    D. Neither A nor B

50. A vehicle with a theft alarm system is being diagnosed for a problem of the alarm will not activate. Technician A says that a misadjusted door could be the cause. Technician B says that a faulty ignition switch could be the cause. Who is correct?

    A. A only
    B. B only
    C. Both A and B
    D. Neither A nor B

PREPARATION EXAM 4

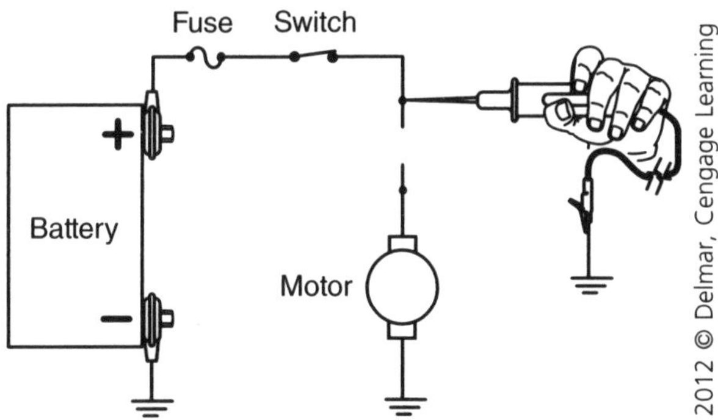

1. Referring to the figure above, Technician A says that the test light will not light because there is an open circuit. Technician B says that the test light should not be used on this type of circuit. Who is correct?

   A. A only
   B. B only
   C. Both A and B
   D. Neither A nor B

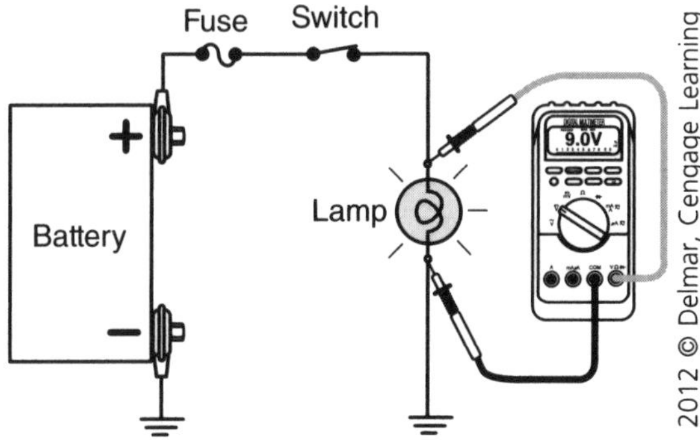

2. The circuit in the figure above is a 12 volt circuit and the battery is fully charged. Which of the following would be most likely to cause the reading?

   A. Bad fuse
   B. Bad switch
   C. Bad lamp
   D. Open circuit at the switch

3. Which tool is recommended by manufacturers to perform voltage measurements on circuits that are controlled or monitored by a control module?

   A. DMM
   B. Test light
   C. Continuity tester
   D. Oscilloscope

4. All of the following conditions will cause reduced current flow EXCEPT:

   A. Loose ground connection
   B. Shorted motor winding
   C. Corroded terminal
   D. Burnt connector

5. Technician A says that an open switch should have infinite resistance. Technician B says that a closed switch should have continuity. Who is correct?

   A. A only
   B. B only
   C. Both A and B
   D. Neither A nor B

6. Technician A says that power must be turned off in the circuit before using an ohmmeter to make a measurement. Technician B says that the ohmmeter applies a small amount of voltage into to the circuit to calculate resistance. Who is correct?

   A. A only
   B. B only
   C. Both A and B
   D. Neither A nor B

7. Viewing waveforms on an oscilloscope would be useful for all of the following conditions EXCEPT:

   A. Checking for intermittent problems in a permanent magnet (PM) generator
   B. Checking a relay coil for a voltage spike
   C. Measuring the voltage drop across a motor
   D. Checking for excessive voltage spikes coming from the generator

8. Technician A says that a scan tool can retrieve codes from the body control module (BCM) on late model vehicles. Technician B says that a scan tool can display live data from some electronic systems. Who is correct?

   A. A only
   B. B only
   C. Both A and B
   D. Neither A nor B

9. Technician A says that a short to ground before the load will cause the circuit protection device to open when the hot-side switch is turned on. Technician B says that a short to ground after the load will cause the circuit protection device to open when the hot-side switch is turned on. Who is correct?

   A. A only
   B. B only
   C. Both A and B
   D. Neither A nor B

10. A vehicle is being diagnosed for a dead battery. During the diagnosis, an excessive parasitic draw is found. Technician A says that removing all of the fuses at once is an effective method of locating the unwanted draw. Technician B says that a "stuck open" trunk light switch could be the cause. Who is correct?

    A. A only
    B. B only
    C. Both A and B
    D. Neither A nor B

11. Technician A says that a fusible link will always burn into two pieces when it blows. Technician B says that a fusible link can be tested with a voltmeter without disconnecting it from the vehicle. Who is correct?

    A. A only
    B. B only
    C. Both A and B
    D. Neither A nor B

12. Which of the following details would be the most likely item to be located on a wiring diagram?

    A. The power and ground distribution for the circuit
    B. The location of the ground connection
    C. Updated factory information about pattern failures
    D. A flowchart for troubleshooting an electrical problem

13. A wiring schematic is being used to troubleshoot an electrical problem. Technician A says that most schematics show the colors of the wires. Technician B says that most schematics are drawn with the power coming from the top of the picture. Who is correct?

    A. A only
    B. B only
    C. Both A and B
    D. Neither A nor B

14. Which tool is the LEAST LIKELY choice to use in the diagnosis of a problem on the data bus network?

    A. Oscilloscope
    B. Digital multimeter
    C. Continuity tester
    D. Scan tool

15. Technician A says that a 12 volt battery that has 6 volts at the posts is 50 percent charged. Technician B says that a 12 volt battery that has 12.6 volts at the posts is overcharged. Who is correct?

    A. A only
    B. B only
    C. Both A and B
    D. Neither A nor B

16. A battery load test has been performed on an automotive battery. The voltage at the end of the 15 second test was 8.4 volts. Technician A says that the battery should be recharged for 15 minutes and then retested. Technician B says that the battery terminals sometimes get warm during this test. Who is correct?

    A. A only
    B. B only
    C. Both A and B
    D. Neither A nor B

17. Technician A says that overcharging a battery will not cause significant long-term damage. Technician B says that in hot weather, more current is needed to charge a battery. Who is right?

    A. A only
    B. B only
    C. Both A and B
    D. Neither A nor B

18. A vehicle is being jumpstarted. Technician A says the engine should be running on the boost vehicle before attempting to crank the dead vehicle. Technician B says the engine should be off while connecting the booster cables. Who is correct?

    A. A only
    B. B only
    C. Both A and B
    D. Neither A nor B

19. A starter current draw test was performed on a late model vehicle. Technician A says that worn starter bushings will cause high current draw. Technician B says that battery terminal corrosion will cause lower than normal current draw. Who is correct?

    A. A only
    B. B only
    C. Both A and B
    D. Neither A nor B

20. A vehicle is being diagnosed for an inoperative starter. A voltage drop test was performed on the solenoid "load side" while the ignition switch is held in the crank mode and .2 volts measured. Technician A says that the solenoid is faulty. Technician B says that this test should only be performed if the battery is at least 75 percent charged. Who is correct?

    A. A only
    B. B only
    C. Both A and B
    D. Neither A nor B

21. A vehicle will not crank and the technician notices that the interior lights do not dim when the ignition switch is moved to the start position. The most likely cause would be which of the following?

    A. Stuck closed starter relay
    B. Loose battery cable connections
    C. Starter mounting bolts loose
    D. Stuck open ignition switch

22. Which of the following functions would be LEAST LIKELY to be performed by the starter solenoid?

    A. Prevents the armature from over-spinning
    B. Push the drive gear out to the flywheel
    C. Connects the "bat" terminal to the "motor" terminal
    D. Provides a path for high current to flow

23. Technician A says that the replacement starter assembly should be inspected carefully prior to installing on the engine. Technician B says that the replacement starter should be bench-tested prior to installing on the engine. Who is correct?

    A. A only
    B. B only
    C. Both A and B
    D. Neither A nor B

24. A maintenance-free battery is low on electrolyte. Technician A says a defective voltage regulator may cause this problem. Technician B says a loose alternator belt may cause this problem. Who is right?

    A. A only
    B. B only
    C. Both A and B
    D. Neither A nor B

25. Which of the following options would the most likely maximum charging voltage on a late-model vehicle?

    A. 13.4 volts
    B. 15.8 volts
    C. 15.6 volts
    D. 14.6 volts

26. A late-model vehicle is being diagnosed for a charging problem. The generator only charges at 12.2 volts. Technician A says that the voltage drop should be checked on the charging output wire. Technician B says that the voltage drop should be checked on the charging ground circuit. Who is correct?

    A. A only
    B. B only
    C. Both A and B
    D. Neither A nor B

27. During an output test of the charging system, a technician finds that the charging current is at 75 amps. The specification for the vehicle is 115 amps. What is the most likely cause?

    A. The vehicle battery has the wrong CCA.
    B. The output test was performed at idle.
    C. The battery cables are oversize.
    D. The voltage regulator has stuck at 100 percent.

28. A PCM controlled charging system is being diagnosed for a low charging problem. Technician A says that a bi-directional scan tool can be used to test the voltage regulation function for fault codes. Technician B says that the alternator can be commanded to charge at full capacity with a bi-directional scan tool. Who is correct?

    A. A only
    B. B only
    C. Both A and B
    D. Neither A nor B

29. An inoperative lamp is being diagnosed. Technician A says that an open wire between the switch and the bulb could be the cause. Technician B says that a shorted switch could be the cause. Who is correct?

    A. A only
    B. B only
    C. Both A and B
    D. Neither A nor B

30. The headlights work on high beams but are inoperative on low beams. Technician A says that the dimmer switch could be the fault. Technician B says that both bulbs could have open low beam filaments. Who is correct?

    A. A only
    B. B only
    C. Both A and B
    D. Neither A nor B

31. The trunk light stays on continuously on a late model vehicle. Technician A says that an open trunk light switch is the likely cause. Technician B says that a shorted wire near the trunk light switch could be the cause. Who is correct?

    A. A only
    B. B only
    C. Both A and B
    D. Neither A nor B

32. Technician A says that many brake light switches are adjustable. Technician B says that many brake light switches also have circuits that disengage the cruise control. Who is correct?

    A. A only
    B. B only
    C. Both A and B
    D. Neither A nor B

33. Technician A says that the turn signal flasher is not load sensitive. Technician B says that the hazard light flasher is load sensitive. Who is correct?

    A. A only
    B. B only
    C. Both A and B
    D. Neither A nor B

34. The bulbs for the back-up lights have both failed. Technician A installs some dielectric grease on the terminals of the replacement bulb before installing. Technician B tests the voltage available to the back-up lights after replacing the bulbs. Who is correct?

A. A only
B. B only
C. Both A and B
D. Neither A nor B

35. The temperature sending unit needs to be replaced on a late-model vehicle. Technician A says that the whole cooling system will need to be drained during this process. Technician B says that the engine should be cooled down prior to performing this repair. Who is correct?

A. A only
B. B only
C. Both A and B
D. Neither A nor B

36. Technician A says that a corroded connection at the fuel sending unit could cause a electro-magnetic fuel gauge to read too high. Technician B says that a grounded connection at the fuel sending unit could cause a bi-metallic fuel gauge to read high. Who is correct?

A. A only
B. B only
C. Both A and B
D. Neither A nor B

37. The fuel gauge reads full at all times. All of the other instrument panel gauges work correctly. Technician A says that the fuel sending unit wire could be broken. Technician B says that the gauges fuse could be blown. Who is correct?

A. A only
B. B only
C. Both A and B
D. Neither A nor B

38. Warning lights and warning devices are generally activated by which of the following components?

A. An ISO relay
B. Closing of a switch or sensor
C. Opening of a switch or sensor
D. An inertia switch

39. The indicator for the bright lights does not work but the bright and dim headlights function correctly. Technician A says that the bulb for the bright indicator could be defective. Technician B says that the dimmer switch could be defective. Who is correct?

A. A only
B. B only
C. Both A and B
D. Neither A nor B

40. The chime reminder for the headlights sounds at times as the vehicle is driven over rough roads. Which of the following conditions would be the most likely cause for this problem?

    A. Open wire in the headlight circuit
    B. Faulty headlight bulb
    C. Park light bulb burnt out
    D. Headlight wire rubbing an instrument panel bracket

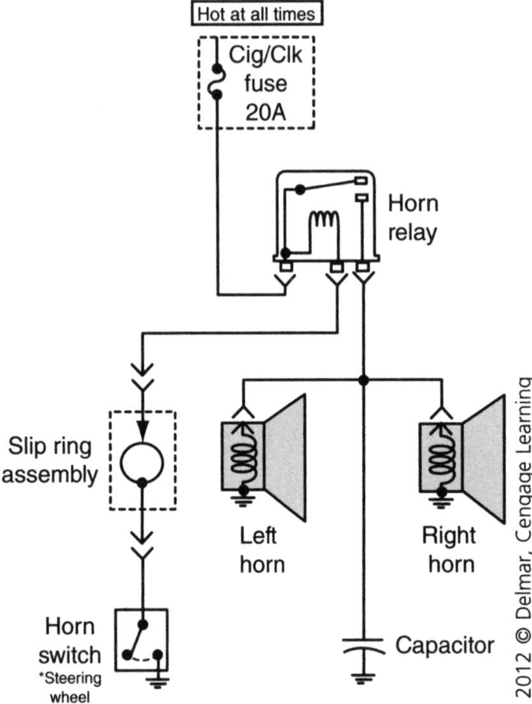

41. Referring to the figure above, Technician A says that the "Cig/Clk" fuse powers both sides of the horn relay. Technician B says that horn switch provides voltage to the relay coil to when the switch is depressed. Who is correct?

    A. A only
    B. B only
    C. Both A and B
    D. Neither A nor B

42. What is the most likely cause of a windshield wiper system that only works on low speed?

    A. A blown fuse
    B. A faulty multi-function switch
    C. A lose ground at the wiper motor
    D. An open park switch

43. The windshield washer system does not spray any water on the windshield when the switch is depressed. A voltage test was performed at the pump motor while the switch was depressed and 13 volts was measured. What is the most likely cause of this problem?

A. A faulty washer pump switch
B. A faulty washer pump relay
C. An open wire in the washer pump circuit
D. A faulty washer pump motor

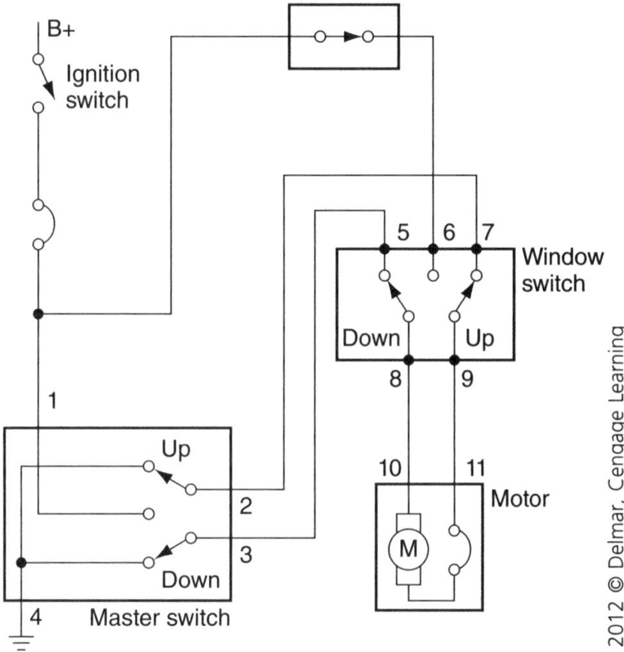

44. Referring to the figure above, the power window motor operates correctly from the master switch but does not operate at all from the window switch. Technician A says that the built-in circuit breaker in the motor could be defective. Technician B says that terminal #6 at the window switch could be broken. Who is correct?

A. A only
B. B only
C. Both A and B
D. Neither A nor B

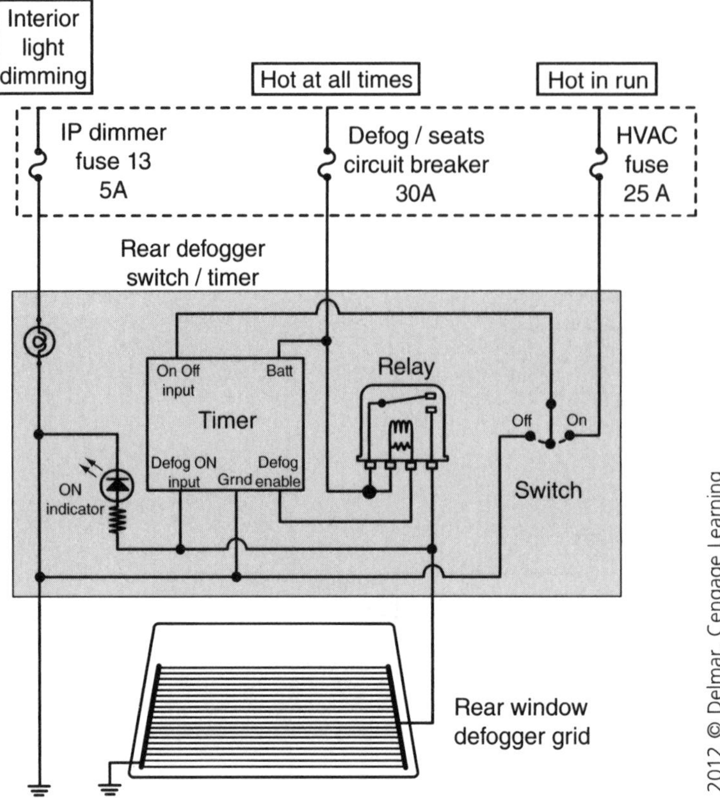

45. Referring to the figure above, Technician A says that the HVAC fuse supplies the "load side" of the defogger relay. Technician B says that the HVAC fuse is hot at all times. Who is correct?

A. A only
B. B only
C. Both A and B
D. Neither A nor B

46. An inoperative sunroof is being diagnosed. A blown fuse for the sunroof is found. Technician A says the cables should be checked for a binding problem. Technician B says that the motor should be checked for a shorting problem with an ohmmeter. Who is correct?

A. A only
B. B only
C. Both A and B
D. Neither A nor B

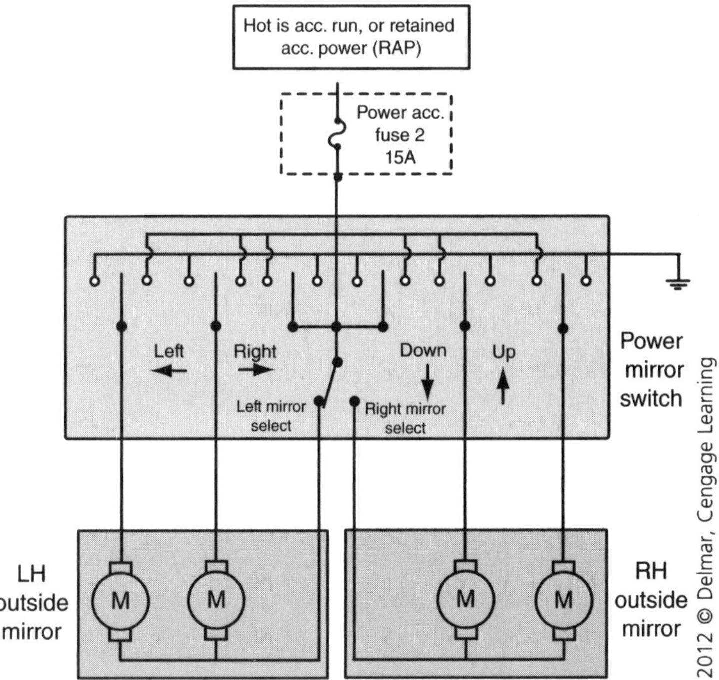

47. Referring to the figure above, the left mirror functions normally but the right side power mirror does not function at all. Technician A says that mirror select switch could be defective. Technician B says that the power mirror switch could have a disconnected ground. Who is correct?

   A. A only
   B. B only
   C. Both A and B
   D. Neither A nor B

48. A radio assembly is being replaced on a late-model vehicle. Technician A says that the radio fuse should be removed for 30 minutes prior to removing the radio. Technician B says that the anti-theft unlock code has to be programmed into the radio before it will function. Who is correct?

   A. A only
   B. B only
   C. Both A and B
   D. Neither A nor B

49. A late-model vehicle is being diagnosed with an anti-theft problem. Technician A says that the system can usually be overridden by using a jumper wire at the data link connector (DLC). Technician B says that a scan tool can be used to retrieve fault codes from the anti-theft system. Who is correct?

   A. A only
   B. B only
   C. Both A and B
   D. Neither A nor B

50. A digital clock loses the correct time each time the ignition is turned off. The most likely cause of this problem is:

 A. Faulty body control module (BCM)
 B. Blown "keep alive memory" fuse
 C. Faulty gateway module
 D. Blown radio fuse

PREPARATION EXAM 5

1. A late-model vehicle is being diagnosed for an inoperative heated steering wheel. Technician A says that a faulty clock spring could cause this problem. Technician B says that the steering wheel heater grid can be replaced without removing the steering wheel. Who is correct?

 A. A only
 B. B only
 C. Both A and B
 D. Neither A nor B

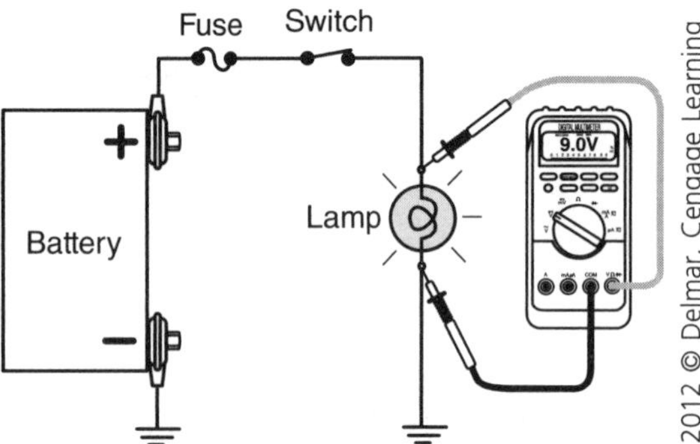

2. The circuit in the figure above is a 12 volt circuit and the battery is fully charged. Technician A says that the switch could have burnt contacts. Technician B says that the circuit could have a poor ground connection. Who is correct?

 A. A only
 B. B only
 C. Both A and B
 D. Neither A nor B

3. Technician A says that high current draw can be caused by dirty electrical connections. Technician B says that low current draw can be caused by corroded electrical connections. Who is correct?

 A. A only
 B. B only
 C. Both A and B
 D. Neither A nor B

4. Technician A says that burnt electrical contacts will decrease the electrical resistance in a circuit. Technician B says that an open switch should have continuity. Who is correct?
   A. A only
   B. B only
   C. Both A and B
   D. Neither A nor B

5. Which of the following tests would be most likely performed with an oscilloscope?
   A. Inspecting a Permanent Magnet (PM) generator for a chipped tooth on the reluctor ring
   B. Testing the voltage drop in the positive battery cable
   C. Inspecting a relay coil resistance
   D. Testing the voltage drop on the charging output circuit

6. Technician A says that a scan tool can retrieve diagnostic trouble codes from the airbag control module on late-model vehicles. Technician B says that a scan tool can display live data from the body control module. Who is correct?
   A. A only
   B. B only
   C. Both A and B
   D. Neither A nor B

7. A fused jumper wire can be used for all of the following procedures EXCEPT:
   A. Bypassing the dimmer switch during diagnosis
   B. Bypassing the rear defogger relay during diagnosis
   C. Bypassing the blower motor relay during diagnosis
   D. Bypassing a potentiometer reference voltage to ground during diagnosis

8. Technician A says that a circuit with corrosion in the wiring will create a short circuit. Technician B says that a power wire that rubs a metal surface for a long period of time could create a short to ground. Who is correct?
   A. A only
   B. B only
   C. Both A and B
   D. Neither A nor B

9. Technician A says that an amp clamp is a useful tool to use when checking for key-off drain. Technician B says that a faulty rectifier bridge in the alternator could cause excessive key-off drain. Who is correct?
   A. A only
   B. B only
   C. Both A and B
   D. Neither A nor B

10. Technician A says that circuit breakers are solid-state components. Technician B says that some circuit breakers automatically reset when they cool down. Who is correct?
    A. A only
    B. B only
    C. Both A and B
    D. Neither A nor B

11. Technician A says that wiring diagrams give the location of each splice used in the circuit. Technician B says that wiring diagrams use electrical symbols to represent the electrical components. Who is correct?

    A. A only
    B. B only
    C. Both A and B
    D. Neither A nor B

12. A vehicle with a data bus network problem is being diagnosed. Technician A says that the owner will not likely notice any unusual problems if the data bus wires become shorted together. Technician B says data bus communication happens when voltage pulses are sent from module to module thousands of times per second. Who is correct?

    A. A only
    B. B only
    C. Both A and B
    D. Neither A nor B

13. All of the following methods are possible methods of determining the state-of-charge of an automotive battery EXCEPT:

    A. Testing the specific gravity
    B. Load test
    C. Open circuit voltage test
    D. Digital battery test

14. A digital battery tester would be most likely used for which of the following purposes?

    A. Measuring resistance of control modules
    B. Testing starter current draw
    C. Measuring voltage of control modules
    D. Testing the impedance of the battery

15. A battery load test has been performed on an automotive battery. The voltage at the end of the 15 second test was 10.4 volts. Technician A says that the battery should be recharged for 15 minutes and then retested. Technician B says that the connectors of the test leads sometimes get warm during this test. Who is correct?

    A. A only
    B. B only
    C. Both A and B
    D. Neither A nor B

16. The battery housing received some damage from driving a vehicle on rough roads. Electrolyte spilled all over the battery tray. Technician A says that brake cleaner should be used to clean the area. Technician B says that baking soda could be used to neutralize the battery acid. Who is correct?

    A. A only
    B. B only
    C. Both A and B
    D. Neither A nor B

17. A starter current draw test was performed on a late model vehicle. Technician A says that worn starter brushes will cause high current draw. Technician B says that battery terminal corrosion will cause higher than normal current draw. Who is correct?

A. A only
B. B only
C. Both A and B
D. Neither A nor B

18. A vehicle is being diagnosed for an inoperative starter. A voltage drop test was performed on the solenoid "load side" while ignition switch is held in the crank mode, and 3 volts is measured. Technician A says that the solenoid is faulty. Technician B says that this test should only be performed if the battery is at least 75 percent charged. Who is correct?

A. A only
B. B only
C. Both A and B
D. Neither A nor B

19. The starter solenoid performs all of the following functions EXCEPT:

A. Provides a path for high current to flow into the starter
B. Push the drive gear out to the flywheel
C. Connects the "bat" terminal to the "motor" terminal
D. Provides gear reduction to increase torque in the starter

20. All of these conditions would cause the starter not to crank the engine **EXCEPT:**

A. The battery is not connected to the starter motor.
B. The solenoid does not engage the starter drive pinion with the engine flywheel.
C. Failure of the control circuit to switch the large-current circuit.
D. The starter drive pinion fails to disengage from the flywheel.

21. All of the following conditions could cause the generator to have zero output EXCEPT:

A. Blown fusible link in charge circuit
B. An open rotor coil
C. A full-fielded rotor
D. A faulty voltage regulator

22. An alternator with a 120-ampere rating produces 85 amps during an output test. A V-belt drives the alternator and the belt is at the specified tension. Technician A says the V-belt may be worn and bottomed in the pulley. Technician B says the alternator pulley may be misaligned with the crankshaft pulley. Who is right?

A. A only
B. B only
C. Both A and B
D. Neither A nor B

23. Which of the following options would be the LEAST LIKELY maximum charging voltage on a late model vehicle?

    A. 13. 9 volts
    B. 15.8 volts
    C. 14. 3 volts
    D. 14.6 volts

24. A late model vehicle is being diagnosed for a charging problem. The generator only charges at 12.4 volts. A voltage drop test is performed on the charging output wire and 1.8 volts is measured. Technician A says that a blown fusible link in the output circuit could be the cause. Technician B says that a loose nut at the charging output connector could be the cause. Who is correct?

    A. A only
    B. B only
    C. Both A and B
    D. Neither A nor B

25. The generator needs to be replaced on a late model vehicle. Technician A says that the negative battery cable should be removed during this process. Technician B says the drive belt should be closely inspected during this process. Who is correct?

    A. A only
    B. B only
    C. Both A and B
    D. Neither A nor B

26. Technician A says that a corroded parking lamp socket could cause dimmer than normal lamp operation. Technician B says that a single filament bulb can be used in place of a dual filament bulb. Who is correct?

    A. A only
    B. B only
    C. Both A and B
    D. Neither A nor B

27. Which of the following electrical components would be most likely used to control the brightness of the instrument panel lights?

    A. Photo resistor
    B. Transducer
    C. Piezo resistor
    D. Rheostat

28. The under-hood light stays on continuously on a late model vehicle. Technician A says that an open under-hood light switch is the likely cause. Technician B says that a shorted wire near the under-hood light switch could be the cause. Who is correct?

    A. A only
    B. B only
    C. Both A and B
    D. Neither A nor B

29. Technician A says that most brake light switches are located near the brake pedal. Technician B says that many brake light switches also have circuits that activate the shift interlock system. Who is correct?

    A. A only
    B. B only
    C. Both A and B
    D. Neither A nor B

30. The right turn signal indicator comes on when the park lights are turned on. Technician A says that a bad ground at the right front park and turn signal lamp socket could be the cause. Technician B says that a single filament bulb in a dual filament socket could be the cause. Who is correct?

    A. A only
    B. B only
    C. Both A and B
    D. Neither A nor B

31. Inoperative back-up lights are being diagnosed. Technician A says that a faulty multi-function switch could be the cause. Technician B says that a stuck open back-up lamp switch could be the cause. Who is correct?

    A. A only
    B. B only
    C. Both A and B
    D. Neither A nor B

32. The oil pressure gauge intermittently moves to the area above the high setting. Technician A says that the oil pressure should be checked with a manual gauge to verify oil pressure. Technician says that an open wire in the oil gauge could be the cause of this problem. Who is correct?

    A. A only
    B. B only
    C. Both A and B
    D. Neither A nor B

33. A vehicle with electronic instrumentation has gauge accuracy problems. Which of the following actions would be the most likely method to diagnose the fault?

    A. Swap the panel with a like-new one.
    B. Replace the suspect sender unit(s).
    C. Ground the sender wire at the suspect sender unit(s).
    D. Connect a scan tool and compare display to gauge readings.

34. The fuel gauge reads empty at all times. All of the other instrument panel gauges work correctly. Technician A says that the fuel sending unit wire could be shorted to ground. Technician B says that the gauges fuse could be blown. Who is correct?

A. A only

B. B only

C. Both A and B

D. Neither A nor B

35. The temperature sending unit needs to be replaced on a late model vehicle. Technician A says that this repair can be made without draining the whole cooling system. Technician B says that the instrument cluster should be disconnected during this process. Who is correct?

A. A only

B. B only

C. Both A and B

D. Neither A nor B

36. The indicator for the bright lights does not work but the bright and dim headlights function correctly. Which of the following conditions would be the most likely cause of this problem?

A. Headlight switch

B. Faulty bulb socket for the bright indicator

C. Dimmer switch

D. Multi-function switch

37. The door ajar chime sounds at times at the vehicle is driven over rough roads. Technician A says that a misadjusted door striker could be the cause. Technician B says that a wire for one of the doors could be rubbing a metal bracket. Who is correct?

A. A only

B. B only

C. Both A and B

D. Neither A nor B

38. The horn blows intermittently on a late-model vehicle. Which of the following conditions would be the LEAST LIKELY cause of this problem?

A. Wire rubbing a ground near the base of the steering column

B. Faulty clock spring

C. Faulty horn switch

D. Horn

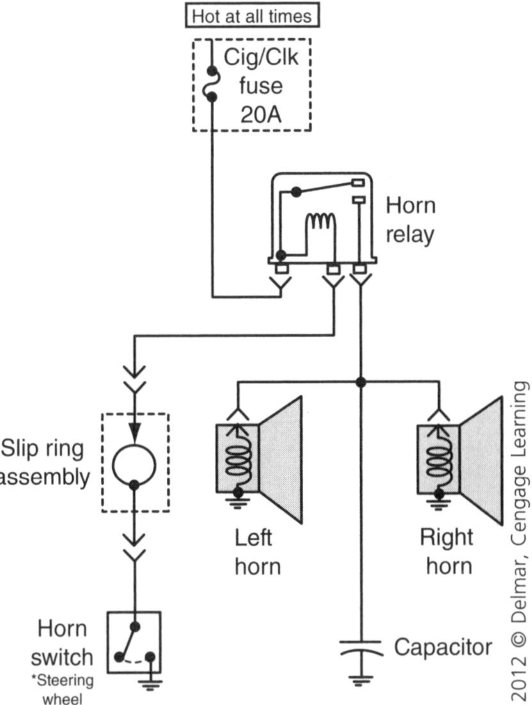

39. Referring to the figure above, Technician A says that the horns are wired in series with each other. Technician B says that the horn switch provides ground to the relay coil when the switch is depressed. Who is correct?

A. A only
B. B only
C. Both A and B
D. Neither A nor B

40. The windshield washer motor is inoperative on a late model vehicle. A voltage test was performed at the pump motor while the washer switch was depressed and 0 volts was measured. What is the most likely cause of this problem?

A. A shorted washer pump switch
B. A shorted park switch
C. An open wire in the washer pump circuit
D. A faulty washer pump motor

41. The power windows work from the master switch but do not work from any of the other switches in the vehicle. Technician A says that the lockout switch could be stuck closed. Technician B says that the master switch main ground connection could be disconnected. Who is correct?

A. A only
B. B only
C. Both A and B
D. Neither A nor B

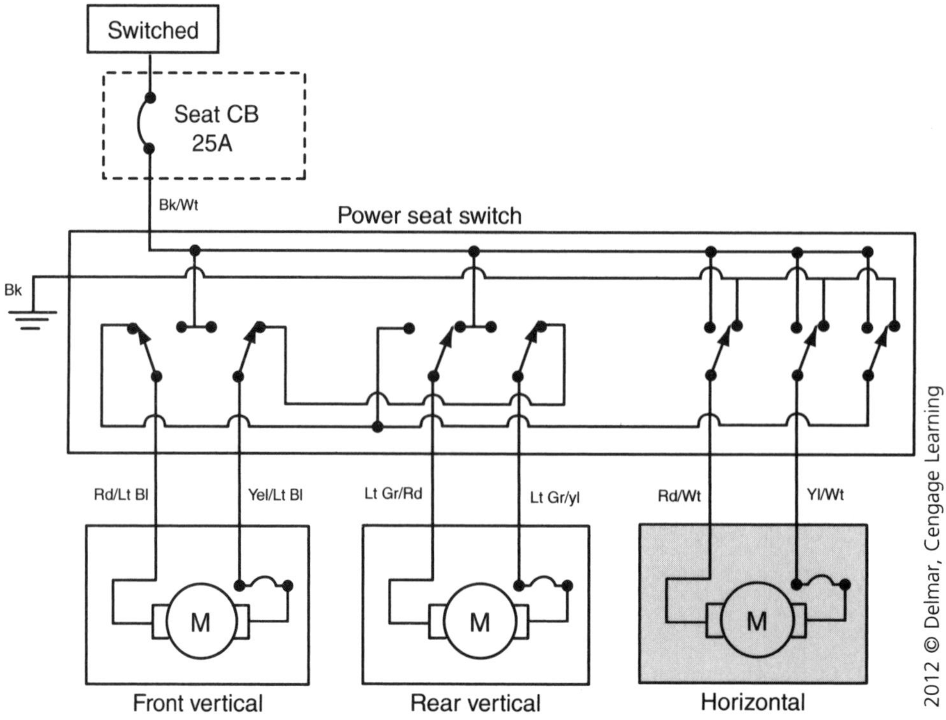

42. Referring to the figure above, the horizontal seat motor is inoperative. Technician A says that the Rd/Wt wire could be broken. Technician B says that power seat switch could have a broken contact. Who is correct?

A. A only
B. B only
C. Both A and B
D. Neither A nor B

43. The rear defogger grid clears in all areas of the rear window except for one section. Technician A says that a voltmeter should be used to check for a bad ground connection. Technician B says that a voltmeter should be used to check for an open circuit in the grid. Who is correct?

A. A only
B. B only
C. Both A and B
D. Neither A nor B

44. The rear defogger does not clear the glass at all when in operation. Technician A says that the rear defogger relay could be defective. Technician B says that one of the heat strips could have a break in it. Who is correct?

A. A only
B. B only
C. Both A and B
D. Neither A nor B

45. A radio assembly is being replaced on a late-model vehicle. Technician A says that the anti-theft unlock code may have to be programmed into the radio before it will function. Technician B says that care should be taken to prevent static electricity when handling the radio assembly. Who is correct?

A. A only
B. B only
C. Both A and B
D. Neither A nor B

46. Technician A says that a radio antenna can be checked with an ammeter. Technician B says that the radio antenna coaxial cable can be checked with an ohmmeter. Who is correct?

A. A only
B. B only
C. Both A and B
D. Neither A nor B

47. The cruise control is inoperative on a late-model vehicle. Technician A says that a scan tool can be used to troubleshoot many of the cruise control components. Technician B says that a misadjusted brake switch could be the cause. Who is correct?

A. A only
B. B only
C. Both A and B
D. Neither A nor B

48. A late-model vehicle is being diagnosed with an anti-theft problem. Technician A says that many ignition keys have special circuitry that has to be replaced at regular intervals. Technician B says that a scan tool can be used to view sensor data associated with the anti-theft system. Who is correct?

A. A only
B. B only
C. Both A and B
D. Neither A nor B

49. All of the following practices are common when diagnosing supplemental inflatable restraint systems EXCEPT:

A. Measuring resistance of the inflator module
B. Using a scan tool to retrieve trouble codes
C. Using a load tool to check for shorts and opens in the inflator module
D. Removing the negative battery cable prior to working around inflator devices

50. Technician A says that all wire repairs on the starting system must be done with weather resistant methods. Technician B says that the wires should be voltage drop tested after repairs have been made. Who is correct?

A. A only
B. B only
C. Both A and B
D. Neither A nor B

PREPARATION EXAM 6

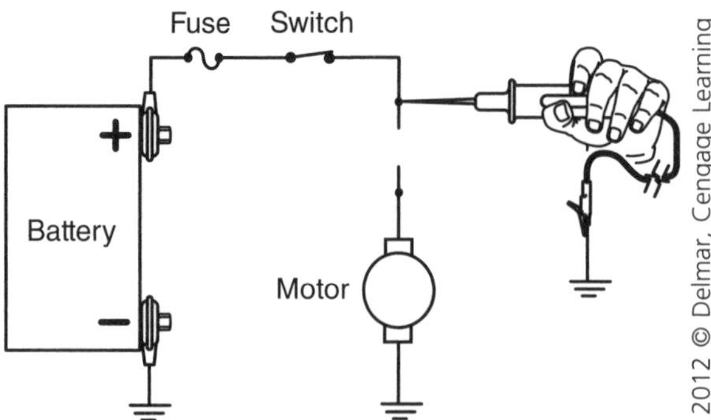

1. Referring to the figure above, Technician A says that the test light will light because it is connected before the open circuit. Technician B says that the test light would be dim because of the open circuit. Who is correct?

   A. A only
   B. B only
   C. Both A and B
   D. Neither A nor B

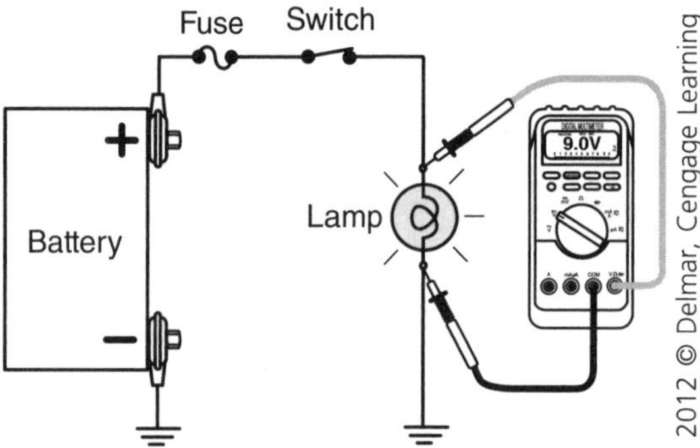

2. The circuit in the figure above is a 12 volt circuit and the battery is fully charged. Technician A says that the fuse panel could have burnt contacts. Technician B says that the bulb could be open. Who is correct?

   A. A only
   B. B only
   C. Both A and B
   D. Neither A nor B

3. Technician A says that a high impedance meter should be used when measuring voltages in electronic circuits. Technician B says that a digital meter with at least 1,000 megohm capacity should be used on electronic circuits. Who is correct?

   A. A only
   B. B only
   C. Both A and B
   D. Neither A nor B

4. Technician A says that using an amp clamp to measure current flow in an electrical circuit will interrupt the circuit. Technician B says that some low amp probes can measure milliamps with accuracy. Who is correct?

   A. A only
   B. B only
   C. Both A and B
   D. Neither A nor B

5. Technician A says that an electrical switch that has continuity will allow current to flow when the switch is closed. Technician B says that a piece of wire that has high resistance will have increased current flow. Who is correct?

   A. A only
   B. B only
   C. Both A and B
   D. Neither A nor B

6. Technician A says that an oscilloscope can be used to view a faulty signal from a potentiometer. Technician B says that an oscilloscope can be used to view a glitch created by a permanent magnet (PM) generator. Who is correct?

   A. A only
   B. B only
   C. Both A and B
   D. Neither A nor B

7. A vehicle is in the repair shop with an inoperative power trunk release system. A scan tool is connected and the power trunk release operates when a functional test is performed. Technician A says that a faulty trunk release switch could be the problem. Technician B says that a poor connection at the trunk release motor could be the problem. Who is correct?

   A. A only
   B. B only
   C. Both A and B
   D. Neither A nor B

8. Which of the following tests would be LEAST LIKELY done by using a fused jumper wire?

   A. Supplying power to a blower motor
   B. Reference voltage supplied to a speed sensor
   C. Bypassing a horn relay
   D. Supplying power to a power window motor

9. A vehicle is in the repair shop with an inoperative power seat. During diagnosis, a blown power seat fuse is located. Technician A says that a short to ground between the motor and the motor ground could be the cause. Technician B says that a binding power seat motor could be the cause. Who is correct?

   A. A only
   B. B only
   C. Both A and B
   D. Neither A nor B

10. Technician A says that a broken wire to a door jam switch will likely cause a parasitic draw problem. Technician B says that a blown horn fuse will likely cause a parasitic draw problem Who is correct?

    A. A only
    B. B only
    C. Both A and B
    D. Neither A nor B

11. Technician A says that a fusible link will sometimes burn into two pieces when a short circuit occurs. Technician B says that a fusible link can be tested with an ammeter without disconnecting it from the vehicle. Who is correct?

    A. A only
    B. B only
    C. Both A and B
    D. Neither A nor B

12. Technician A says that a wiring diagram uses schematic symbols to represent electrical components. Technician B says that wiring diagrams usually show the splice and connector numbers for each circuit. Who is correct?

    A. A only
    B. B only
    C. Both A and B
    D. Neither A nor B

13. All of the following electrical tools could be used to diagnose a data bus network problem EXCEPT:

    A. Oscilloscope
    B. Digital multimeter
    C. Analog voltmeter
    D. Scan tool

14. Technician A says that a 12 volt battery that has 12.2 volts at the posts is 50 percent charged. Technician B says that a 12 volt battery that has 12.6 volts at the posts is fully charged. Who is correct?

    A. A only
    B. B only
    C. Both A and B
    D. Neither A nor B

15. A battery load test has been performed on an automotive battery. The voltage at the end of the 15 second test was 7.5 volts. Technician A says that the battery failed the test and should be replaced. Technician B says that extreme care should be taken when performing this test. Who is correct?

    A. A only
    B. B only
    C. Both A and B
    D. Neither A nor B

16. Technician A says that all replacement batteries should be slow charged for one hour prior to installing into the vehicle. Technician B says that the battery cable terminals should be cleaned and protected when installing a replacement battery. Who is correct?

    A. A only
    B. B only
    C. Both A and B
    D. Neither A nor B

17. Technician A says that eye protection should be worn when jumpstarting a dead battery with a booster battery. Technician B says that the last connection should be at the positive post of the dead battery when jumpstarting. Who is correct?

    A. A only
    B. B only
    C. Both A and B
    D. Neither A nor B

18. An engine is being diagnosed for a slow crank problem. Technician A says that testing the voltage drop on the battery cables while cranking the engine will reveal cable problems. Technician B says that worn starter brushes could cause the slow crank problem. Who is correct?

    A. A only
    B. B only
    C. Both A and B
    D. Neither A nor B

19. A vehicle will not crank and the technician notices that the interior lights get very dim when the ignition switch is moved to the start position. The most likely cause would be which of the following?

    A. Stuck closed starter relay
    B. Loose battery cable connections
    C. Starter mounting bolts loose
    D. Stuck open ignition switch

20. Technician A says that the starter solenoid pull-in winding can be tested by touching the "start" terminal and the "bat" terminal with the ohmmeter leads. Technician B says that the starter solenoid hold-in winding can be tested by touching the "start" terminal and the solenoid case with the ohmmeter leads. Who is correct?

    A. A only
    B. B only
    C. Both A and B
    D. Neither A nor B

21. Which of the following conditions would be LEAST LIKELY to cause a slow crank problem?

    A. A mistimed engine
    B. Worn starter brushes
    C. Battery to with a higher CCA rating than original specifications
    D. An engine with tight main bearings

22. Technician A says that the replacement starter assembly should be bench-tested prior to installing on the engine. Technician B says that the starter connectors and terminals should be inspected and cleaned prior to installing a replacement starter. Who is correct?

    A. A only

    B. B only

    C. Both A and B

    D. Neither A nor B

23. Which of the following would be the most likely condition to cause the generator to overcharge?

    A. Blown fusible link in charge circuit

    B. An open rotor coil

    C. A full-fielded rotor

    D. A faulty voltage regulator

24. A vehicle is being diagnosed for a charging problem. The generator produced 125 amps during the output test and it is rated at 130 amps. Technician A says that the generator could have a bad diode. Technician B says that the generator drive pulley could be too large in diameter. Who is correct?

    A. A only

    B. B only

    C. Both A and B

    D. Neither A nor B

25. A late model vehicle is being diagnosed for a charging problem. The generator only charges at 12.4 volts. Technician A says that the charging output wire may have an excessive voltage drop. Technician B says that the charging ground circuit may have an excessive voltage drop. Who is correct?

    A. A only

    B. B only

    C. Both A and B

    D. Neither A nor B

26. The wiring and connections of the charging system should be checked in all of the following ways during generator replacement EXCEPT:

    A. Inspect the stator resistance.

    B. Inspect the connections for tightness.

    C. Inspect the wire insulation for cuts and cracks.

    D. Inspect the routing of the wires and harnesses.

27. The generator needs to be replaced on a late-model vehicle. Technician A says that the replacement generator should have the same diameter pulley. Technician B says that the replacement generator should be disassembled and tested prior to installing on the vehicle. Who is correct?

    A. A only

    B. B only

    C. Both A and B

    D. Neither A nor B

28. The right side headlight of a vehicle is very dim and the left side headlight is normal. Technician A says that the dimmer switch is likely defective. Technician B says that the left side headlight could have a bad ground. Who is correct?

    A. A only
    B. B only
    C. Both A and B
    D. Neither A nor B

29. Technician A says that the daytime running lights utilize the headlight filaments on most late-model vehicles. Technician B says that the daytime running lights will use less current than the headlights. Who is correct?

    A. A only
    B. B only
    C. Both A and B
    D. Neither A nor B

30. The fog lights are dimmer than normal and a voltage drop test across the bulbs reveals 10 volts. Which of the following conditions would LEAST LIKELY cause this problem?

    A. Faulty fog light bulbs
    B. Faulty fog light switch contacts
    C. Faulty fog light relay
    D. Faulty connection near the fog light switch

31. Which of the following problems would most likely cause the stop lights to be inoperative?

    A. Stuck closed lamp back-up switch
    B. Incorrect turn signal flasher
    C. Blown headlight fuse
    D. Blown stop light fuse

32. A vehicle is being diagnosed with a problem with the parking assist system. Technician A says a scan tool could be used to retrieve trouble codes from the control module. Technician B says that mud on the proximity sensors could cause the system to malfunction. Who is correct?

    A. A only
    B. B only
    C. Both A and B
    D. Neither A nor B

33. Technician A says to use dielectric grease on the trailer wiring connector to reduce the chance of corrosion. Technician B says that the turn signal flasher may flash faster when the vehicle is connected to a trailer. Who is correct?

    A. A only
    B. B only
    C. Both A and B
    D. Neither A nor B

34. The oil pressure gauge intermittently moves to the "low" setting. Technician A says that the oil and filter should be changed. Technician says that a faulty oil sending could be the cause of this problem. Who is correct?

    A. A only
    B. B only
    C. Both A and B
    D. Neither A nor B

35. The temperature sending unit needs to be replaced on a late-model vehicle. Technician A says that this process can be completed without draining the whole cooling system. Technician B says that the engine should be cooled down prior to performing this repair. Who is correct?

    A. A only
    B. B only
    C. Both A and B
    D. Neither A nor B

36. A vehicle is being diagnosed for a problem of all of the gauges being inoperative. The vehicle is equipped with an electronic cluster that has self-diagnostic capabilities. Technician A says that this problem could be diagnosed by using the self-diagnostic function of the instrument cluster. Technician B says that a scan tool should be connected to the data link connector to view the sensor data activity at the instrument cluster. Who is correct?

    A. A only
    B. B only
    C. Both A and B
    D. Neither A nor B

37. All of the following types of electronic devices are used as inputs to electronic gauge assemblies EXCEPT:

    A. Thermistor
    B. Piezo resistor
    C. Rheostat
    D. Actuator

38. The charging light stays on while driving. Technician A says that a grounded wire near the generator could be the cause. Technician B says that a faulty circuit in the instrument cluster could be the cause. Who is correct?

    A. A only
    B. B only
    C. Both A and B
    D. Neither A nor B

39. The chime reminder for the park assist system sounds at times as the vehicle is driven over rough roads. Which of the following conditions would be the most likely cause for this problem?
    A. Open wire in the park assist circuit
    B. A disconnected chime module
    C. Park assist wire rubbing the vehicle body
    D. Park assist sensor unplugged

40. All of the following could cause an inoperative horn EXCEPT:
    A. Open horn relay coil
    B. Open in the clock spring
    C. Stuck closed horn relay
    D. Blown horn fuse

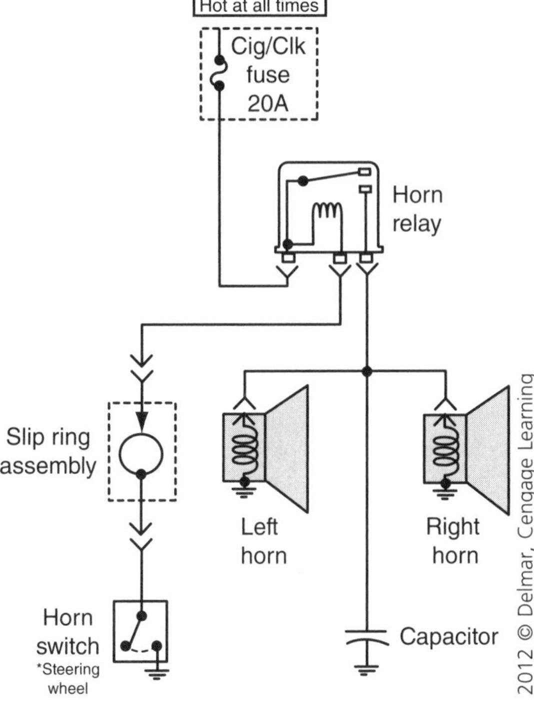

41. Referring to the figure above, all of the following conditions could cause both horns to be inoperative EXCEPT:
    A. Broken wire at the relay output connection
    B. Open "Cig/Clk" fuse
    C. Faulty horn relay
    D. Shorted horn switch

42. The horn blows intermittently while turning the steering wheel on a late-model vehicle. Which of the following conditions would be the most likely cause of this problem?

A. Open horn circuit wire near the power distribution center
B. Faulty clock spring
C. Faulty horn switch
D. Faulty horn

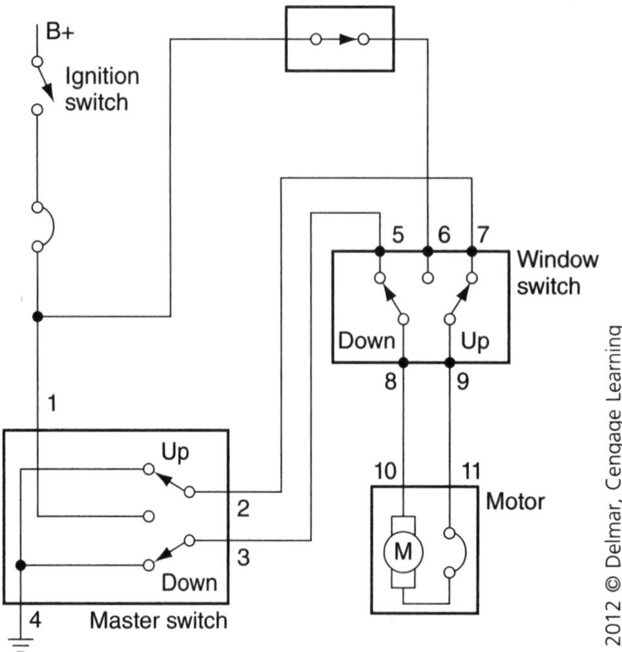

43. Referring to the figure above, the power window motor operates correctly from the window switch but does not operate at all from the master switch. Technician A says that terminal #1 could be broken. Technician B says that terminal #4 at the master switch could be broken. Who is correct?

A. A only
B. B only
C. Both A and B
D. Neither A nor B

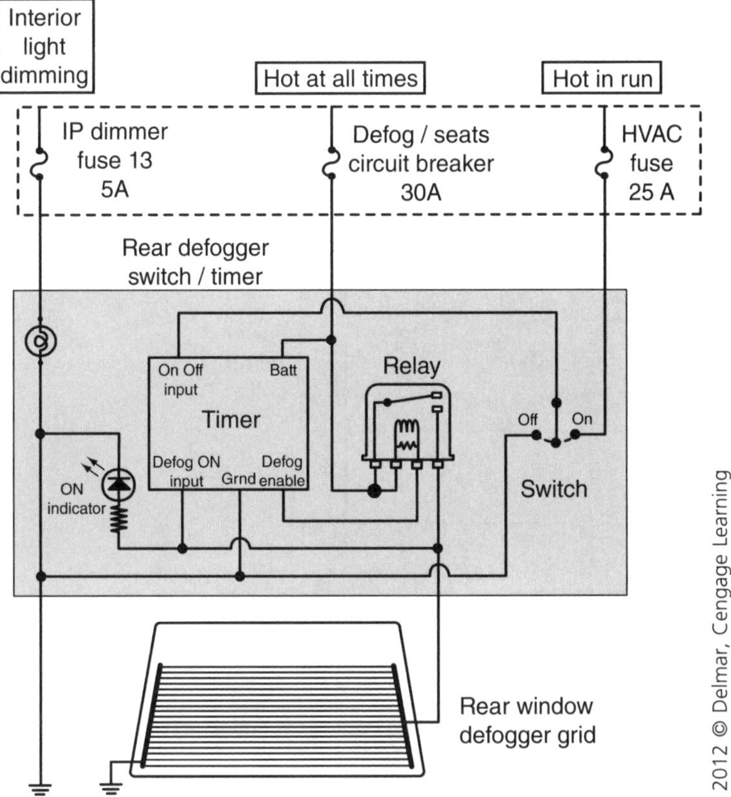

44. Referring to the figure above, the rear defogger does not function but the "on" indicator illuminates when the switch is depressed. Technician A says that the relay coil could be open. Technician B says that there could be an open wire between the relay and the rear defogger grid. Who is correct?

   A. A only
   B. B only
   C. Both A and B
   D. Neither A nor B

45. A minivan with power sliding doors is being diagnosed for a power sliding door problem. The passenger side sliding door does not open with the handle switch, but it works correctly from the driver's switch. Technician A says that the passenger handle switch could be faulty. Technician B says that the passenger handle switch could have a bad connection. Who is correct?

   A. A only
   B. B only
   C. Both A and B
   D. Neither A nor B

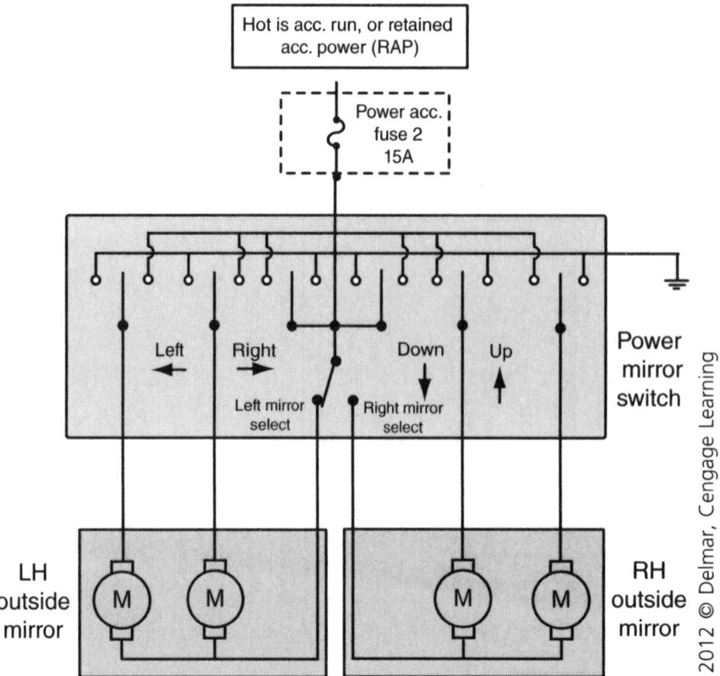

46. Referring to the figure above, the right power mirror functions normally but the left side power mirror does not function in the up and down directions. Technician A says that mirror select switch could be defective. Technician B says that the built-in circuit breaker in the up/downmotor could be defective. Who is correct?

A. A only
B. B only
C. Both A and B
D. Neither A nor B

47. Technician A says that the power antenna uses a bi-directional permanent magnet motor. Technician B says that some power antenna systems use relays to control the up and down circuit operation. Who is correct?

A. A only
B. B only
C. Both A and B
D. Neither A nor B

48. The auxiliary power outlet is inoperative and the fuse is found to be open. What is the LEAST LIKELY cause for this condition?

A. Shorted power wire near the auxiliary connector
B. Foreign metal object in the auxiliary power outlet
C. A faulty electrical device was connected to the outlet
D. Open internal connection at the power outlet

49. Which of the following conditions would be LEAST LIKELY to cause the airbag light to stay on while the vehicle is in operation?

   A. Bad connection at the airbag bulb socket
   B. A shorted airbag inflator
   C. A faulty seatbelt buckle
   D. An open clock spring

50. An airbag system needs to be disarmed during a driver's inflator module replacement. Technician A says that it is necessary to remove the negative battery cable prior to component removal. Technician B says that the inflatable devices should be laid "face down" in a secure area after being removed from a vehicle. Who is correct?

   A. A only
   B. B only
   C. Both A and B
   D. Neither A nor B

SECTION

6 Answer Keys and Explanations

INTRODUCTION

Included in this section are the answer keys for each preparation exam, followed by individual, detailed answer explanations and a reference identifying the designated task area being assessed by each specific question. This additional reference information may prove useful if you need to refer back to the task list located in Section 4 of this book for additional support.

PREPARATION EXAM 1—ANSWER KEY

1. B
2. A
3. A
4. B
5. D
6. A
7. C
8. B
9. C
10. A
11. B
12. D
13. B
14. D
15. A
16. C
17. B
18. B
19. C
20. D
21. D
22. C
23. B
24. D
25. A
26. B
27. D
28. B
29. D
30. B
31. A
32. C
33. C
34. C
35. D
36. C
37. C
38. B
39. A
40. C
41. D
42. C
43. A
44. B
45. B
46. C
47. B
48. B
49. D
50. C

PREPARATION EXAM 1—EXPLANATIONS

1. A control module needs to be programmed following the replacement of the hardware. Technician A says that a multimeter will be needed to measure the resistance of the new software. Technician B says that a module programming database is needed to retrieve the correct program to upload to the new module. Who is correct?

TASK A.11

A. A only
B. B only
C. Both A and B
D. Neither A nor B

Answer A is incorrect. It is never advisable to connect an ohmmeter to a control module. This action would likely damage the module due to voltage being induced from the multimeter.

Answer B is correct. Only Technician B is correct. A module programming database is needed to retrieve the correct program for each module.

Answer C is incorrect. Only Technician B is correct.

Answer D is incorrect. Technician B is correct.

2. Technician A says that voltage drop testing a connector needs to be done when the circuit is energized. Technician B says that voltage drop testing a relay needs to be done when the circuit is de-energized. Who is correct?

TASK A.1

A. A only
B. B only
C. Both A and B
D. Neither A nor B

Answer A is correct. Only Technician A is correct. Performing a voltage drop test is a good way to determine whether a circuit is working correctly. However, this test is useful only when performed on a live circuit.

Answer B is incorrect. All voltage drop tests on electrical components need to be performed when the circuit is energized.

Answer C is incorrect. Only Technician A is correct.

Answer D is incorrect. Technician A is correct.

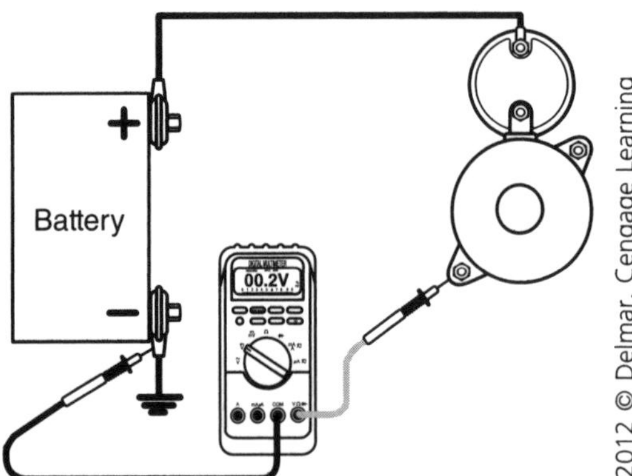

TASK A.1

3. Referring to the figure, Technician A says that the test should be performed while cranking the engine. Technician B says that the reading on the meter indicates that the negative battery cable is faulty. Who is correct?

A. A only
B. B only
C. Both A and B
D. Neither A nor B

Answer A is correct. Only Technician A is correct. All voltage drop tests on electrical components and circuits should be performed while the circuit is energized. The largest amount of current that flows in the battery cables occurs while cranking the engine, so this is the correct time to test the voltage drop on the circuit.

Answer B is incorrect. The reading on the meter shows .2 volts, which is within specifications for most vehicles.

Answer C is incorrect. Only Technician A is correct.

Answer D is incorrect. Technician A is correct.

TASK A.2

4. All of the following conditions can cause reduced current flow in an electrical circuit EXCEPT:

A. Loose terminal connections
B. A hot-side wire that is rubbing a metal component
C. Corrosion inside the wire insulation
D. Wire that is too small

Answer A is incorrect. Loose terminal connections will reduce the current flow because of the added electrical resistance.

Answer B is correct. A hot-side wire that rubs a metal component will cause a large increase in current flow and would likely open a circuit protection device.

Answer C is incorrect. Wire corrosion will reduce the current flow because of the added electrical resistance.

Answer D is incorrect. A wire that is too small will reduce the circuit's ability to carry the correct current because of the limitation of surface area to allow the electrons to flow.

5. Technician A says that an electrical switch that has continuity will have reduced current flow. Technician B says that a piece of wire that has high resistance will have increased current flow. Who is correct?

TASK A.3

A. A only
B. B only
C. Both A and B
D. Neither A nor B

Answer A is incorrect. A switch that has continuity will have normal current flow. A closed switch should have continuity.

Answer B is incorrect. A wire that has high resistance will have reduced current flow due to the increased resistance.

Answer C is incorrect. Neither Technician is correct.

Answer D is correct. Neither Technician is correct. A closed switch should have continuity, which will allow normal current to flow in a circuit. A wire with added electrical resistance will have reduced current flow.

6. Technician A says that intermittent electrical signals can sometimes be diagnosed by using an oscilloscope. Technician B says that a broken tooth on a speed sensor reluctor can be sometimes be diagnosed by using an ohmmeter. Who is correct?

TASK A.4

A. A only
B. B only
C. Both A and B
D. Neither A nor B

Answer A is correct. Only Technician A is correct. An oscilloscope displays voltage over time on a digital screen. Viewing signals this way allows the technician to pick up electrical glitches and intermittent signals that happen very quickly.

Answer B is incorrect. Testing the resistance of a sensor will not pick up something like a broken tooth on a speed sensor. An oscilloscope is an effective tool for finding this type of problem.

Answer C is incorrect. Only Technician A is correct.

Answer D is incorrect. Technician A is correct.

7. A vehicle is in the repair shop with an inoperative horn. A scan tool is connected and the horn operates when a functional test of the horn is performed. Technician A says that a faulty horn switch could be the problem. Technician B says that a faulty clock spring could be the problem. Who is correct?

TASKS A.5, F.9

A. A only
B. B only
C. Both A and B
D. Neither A nor B

Answer A is incorrect. Technician B is also correct.

Answer B is incorrect. Technician A is also correct.

Answer C is correct. Both Technicians are correct. A faulty horn switch as well as a faulty clock spring could be the problem in this question. Using a scan tool to perform an actuator test of the horn proves that the relay has the ability to sound the horn. This test proves that the coil side and the load side of the relay are functioning.

Answer D is incorrect. Both Technicians are correct.

TASK F.2

8. A technician is replacing the electronic module for the park assist system. Which of the following practices would most likely be required to assure a trouble free repair?

   A. Remove the engine noise dampening cover.
   B. Disconnect the negative battery cable prior to beginning the repair.
   C. Remove the bumper retaining fasteners.
   D. Remove the rear hatch lock assembly.

Answer A is incorrect. The engine noise dampening cover would not need to be removed during the replacement of the park assist module.

Answer B is correct. It is advisable to disconnect the negative battery cable prior to replacing the park assist module to prevent voltage spikes from occurring during the procedure.

Answer C is incorrect. The park assist module is not located in the bumper, so the bumper retaining fasteners would not have to be removed.

Answer D is incorrect. The rear hatch lock assembly has nothing to do with the park assist system.

TASKS A.6, F.1

9. A vehicle is in the repair shop with an inoperative power window. During diagnosis, a blown power window fuse is located. Technician A says that a short to ground between the switch and the motor could be the cause. Technician B says that a tight power window motor could be the cause. Who is correct?

   A. A only
   B. B only
   C. Both A and B
   D. Neither A nor B

Answer A is incorrect. Technician B is also correct.

Answer B is incorrect. Technician A is also correct.

Answer C is correct. Both Technicians are correct. A short to ground on the power feed side of the circuit will very likely blow a fuse due to the current quickly rising because the resistance is very low. A binding or tight power window motor can also cause the current flow to rise quickly and blow the power window fuse. Some manufacturers use circuit breakers in the power window circuits for this reason.

Answer D is incorrect. Both Technicians are correct.

TASK A.7

10. A vehicle is in the repair shop for a dead battery. During the diagnosis, an excessive parasitic draw is found. Technician A says that a diode problem in the generator could be the cause. Technician B says that an "open" glove box bulb could be the problem. Who is correct?

   A. A only
   B. B only
   C. Both A and B
   D. Neither A nor B

Answer A is correct. Only Technician A is correct. A shorted diode in the generator could cause an unwanted path for current to flow toward the generator and run the battery down.

Answer B is incorrect. An "open" glove box bulb would never draw any current because electricity does not flow through an open circuit of any type.

Answer C is incorrect. Only Technician A is correct.

Answer D is incorrect. Technician A is correct.

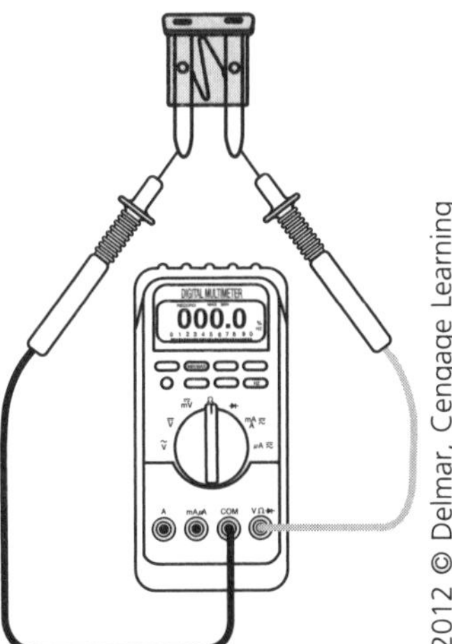

11. Referring to the figure above, Technician A says that the fuse is faulty because it has low resistance. Technician B says that the meter is set to test the resistance of the fuse. Who is correct?

TASK A.8

A. A only

B. B only

C. Both A and B

D. Neither A nor B

Answer A is incorrect. The resistance test in the figure shows a very low reading, which would prove that the fuse was good.

Answer B is correct. Only Technician B is correct. The meter is set to read resistance because the dial is set to the omega symbol.

Answer C is incorrect. Only Technician B is correct.

Answer D is incorrect. Technician B is correct.

12. Which of the following details would be the most likely item to be located on a wiring diagram?

TASK A.9

A. The location of a component

B. A flowchart for troubleshooting an electrical problem

C. The current rating of a circuit

D. The color and circuit number of a wire

Answer A is incorrect. A wiring diagram does not typically give the location of the electrical components.

Answer B is incorrect. A wiring diagram does not typically provide a troubleshooting flowchart. However, an experienced technician can use the wiring diagram to develop a strategy for solving electrical problems.

Answer C is incorrect. A wiring diagram does not typically provide information about the current rating of the circuit. However, an experienced technician can observe the circuit protection devices in the diagram and recognize the limits that cause these devices to open the circuit.

Answer D is correct. Wiring diagrams usually provide the color of each wire as well as the circuit identification number of each wire.

TASK A.11

13. Technician A states that the scan tool receives data from a connector located on the ECM. Technician B states that the scan tool connects to the data bus using data link connector (DLC). Who is correct?

 A. A only
 B. B only
 C. Both A and B
 D. Neither A nor B

Answer A is incorrect. There is no connector at the ECM to connect the scan tool to.

Answer B is correct. Only Technician B is correct. The scan tool connects to the data bus using a data link connector, which is located near the driver's side area. Once connected, the scan tool is a bi-directional device that can read data, retrieve trouble codes and send actuator commands to several systems.

Answer C is incorrect. Only Technician B is correct.

Answer D is incorrect. Technician B is correct.

TASKS B.1, B.2

14. A vehicle is being diagnosed for a battery problem. Technician A says that the battery open circuit voltage should be 12.6 volts before performing a digital battery test. Technician B says that a battery can be accurately load-tested if it has an open circuit voltage of 12 volts. Who is correct?

 A. A only
 B. B only
 C. Both A and B
 D. Neither A nor B

Answer A is incorrect. A battery does not have to be at 12.6 volts in order to perform a digital battery test. Digital battery testers operate by checking the capacitance of the battery and then giving test results, which include the voltage, the cold cranking amps available, as well as whether the battery needs to be charged or replaced.

Answer B is incorrect. A battery needs to be at least 75 percent charged in order to perform a valid load test. The open circuit voltage needs to be at least 12.4 volts to run this test.

Answer C is incorrect. Neither Technician is correct.

Answer D is correct. Neither Technician is correct.

TASK B.3

15. Which of the following tools can be used to maintain electronic memory functions while a battery is disconnected?

 A. A battery jump box with a 12-volt accessory connecting wire
 B. A scan tool
 C. A digital multimeter
 D. A test light

Answer A is correct.A battery jump box typically comes equipped with an accessory plug that can be used to connect a jumper wire into.

Answer B is incorrect. A scan tool does not have the capacity to maintain memory functions while a battery is disconnected.

Answer C is incorrect. A digital multimeter does not have the capacity to maintain memory functions while the vehicle battery is disconnected.

Answer D is incorrect. A test light does not have the capacity to maintain memory functions while the vehicle battery is disconnected.

16. All of the following details are critical when choosing a replacement battery for a vehicle EXCEPT:

TASK B.5

A. Cold cranking amps of the battery
B. Reserve capacity of the battery
C. Battery manufacturer
D. Physical dimensions of the battery case

Answer A is incorrect. The cold cranking amp (CCA) rating is a very important detail when choosing a replacement battery. All replacement batteries should meet the minimum rating set by the vehicle manufacturer.

Answer B is incorrect. The reserve capacity rating is a very important detail when choosing a replacement battery. All replacement batteries should meet the minimum rating set by the vehicle manufacturer.

Answer C is correct. The battery manufacturer is not a critical detail when choosing a replacement battery as long as the other specifications are met.

Answer D is incorrect. It is vital that the replacement battery physically fit into the space provided by the manufacturer.

17. What is the most likely method to be used when professionally cleaning battery cable clamps?

TASK B.5

A. Pneumatic die grinder
B. Wire brush
C. Carburetor cleaner and scraper
D. Screwdriver

Answer A is incorrect. A power tool should never be used around a battery due to the danger of possibly producing sparks as well as having flying debris in this area.

Answer B is correct. A wire brush can safely be used to remove corrosion and residue from the battery terminals and cable ends.

Answer C is incorrect. Carburetor cleaner should never be used near a battery due to the chemical properties which could react to the chemicals associated with the battery.

Answer D is incorrect. A screwdriver would not be an effective tool to use to clean the battery cable clamps.

18. A vehicle with a 4-cylinder engine is being diagnosed for a starting problem. A starter current draw test was performed and the amperage was 295 amps. Technician A says that worn starter brushes could be the problem. Technician B says that a shorted field coil could be the problem. Who is correct?

TASK B.7

A. A only
B. B only
C. Both A and B
D. Neither A nor B

Answer A is incorrect. Worn starter brushes would likely decrease the starter current draw due to the increased electrical resistance.

Answer B is correct. Only Technician B is correct. A shorted field coil could cause elevated starter current draw due to the decrease in electrical resistance.

Answer C is incorrect. Only Technician B is correct.

Answer D is incorrect. Technician B is correct.

TASK B.10

19. A vehicle is in the repair shop for a starting problem. There is a clicking sound at the starter when the key is moved to the start position. Technician A says that the starter solenoid could be the cause. Technician B says that worn starter brushes could be the problem. Who is correct?

    A. A only
    B. B only
    C. Both A and B
    D. Neither A nor B

    Answer A is incorrect. Technician B is also correct.

    Answer B is incorrect. Technician A is also correct.

    Answer C is correct. Both Technicians are correct. A clicking noise while attempting to start an engine could be a faulty starter solenoid or worn starter brushes. A voltage drop test at the solenoid could be performed to further troubleshoot this problem.

    Answer D is incorrect. Both Technicians are correct.

TASK B.9

20. Which of the following steps should the technician do first when beginning to remove a starter from a vehicle?

    A. Pull the starter fuse at the power distribution center.
    B. Remove the starter bolts.
    C. Disconnect the wires at the starter.
    D. Remove the negative battery cable and tape the terminal.

    Answer A is incorrect. It is typically not necessary to remove any fuses while removing a starter from a vehicle.

    Answer B is incorrect. Removing the starter bolts before disconnecting the battery is not advised due to the safety issues. The battery should always be disconnected as the first step.

    Answer C is incorrect. Removing the wires at the starter before disconnecting the battery is a dangerous practice because the positive battery cable directly connecting to the starter. A short to ground could easily occur if the battery is not disconnected first.

    Answer D is correct. The negative battery cable should always be removed prior to disconnecting and removing a starter.

TASK B.8

21. A vehicle is slow to crank. Technician A says that testing the voltage drop on the positive battery cable while cranking the engine will reveal potential problems in the battery. Technician B says that testing the voltage drop on the negative battery cable must be done with the ignition switch in the off position. Who is correct?

    A. A only
    B. B only
    C. Both A and B
    D. Neither A nor B

    Answer A is incorrect. Performing a voltage drop test on the battery cables will test the battery cable, not the battery.

    Answer B is incorrect. Voltage drop tests on any wire or cable needs to be done while the circuit is activated.

    Answer C is incorrect. Neither Technician is correct.

    Answer D is correct. Neither Technician is correct. Testing the voltage drop on the battery cables is done while cranking the engine and the specification is usually .5 volts or less.

22. Technician A says that the starter solenoid pull-in winding can be tested by touching the "start" terminal and the "motor" terminal with the ohmmeter leads. Technician B says that the starter solenoid hold-in winding can be tested by touching the "start" terminal and the solenoid case with the ohmmeter leads. Who is correct?

TASK B.9

A. A only
B. B only
C. Both A and B
D. Neither A nor B

Answer A is incorrect. Technician B is also correct.

Answer B is incorrect. Technician A is also correct.

Answer C is correct. Both Technicians are correct. The starter solenoid windings can be tested with an ohmmeter. The pull-in winding is tested by connecting the ohmmeter leads to the "start" terminal and the "motor" terminal. The hold-in winding can be tested by connecting the ohmmeter leads to the "start" terminal and the solenoid case. Both readings should be very low resistance.

Answer D is incorrect. Both Technicians are correct.

23. All of the following conditions could cause an undercharged battery EXCEPT:

TASK C.1

A. Loose alternator bracket fasteners
B. An oversized charging wire
C. Worn generator brushes
D. Excessive voltage drop in the charging wire

Answer A is incorrect. A loose alternator bracket could cause an unwanted voltage drop in the charging ground circuit.

Answer B is correct. Using a larger charging wire would not cause a charging problem. Larger wire would carry the current easier than a smaller wire.

Answer C is incorrect. Worn generator brushes could reduce the charging output of the generator.

Answer D is incorrect. Excessive voltage drop in the charging wire would reduce the charge rate of the battery.

24. Technician A says that the charging system voltage will operate at approximately 11 to 13 volts. Technician B says that the charging system voltage should be measured while cranking the engine. Who is correct?

TASK C.3

A. A only
B. B only
C. Both A and B
D. Neither A nor B

Answer A is incorrect. The charging voltage should be higher than 11 volts.

Answer B is incorrect. The charging voltage should be measured while the engine is running.

Answer C is incorrect. Neither Technician is correct.

Answer D is correct. Neither Technician is correct. Typical charging voltage should about 13 to 14.5 volts and should be measured while the engine is running.

TASKS
C.4, C.2

25. A vehicle is being diagnosed for a charging problem. The generator produced 85 amps during the output test and it is rated at 135 amps. Technician A says that the generator could have a bad diode. Technician B says that the generator drive pulley could be too small. Who is correct?

A. A only
B. B only
C. Both A and B
D. Neither A nor B

Answer A is correct. Only Technician A is correct. A faulty generator diode could reduce the charging output of the generator. The output should be within 10 percent of the rating of the generator.

Answer B is incorrect. A drive pulley that is smaller would not cause lower charging output due to the faster spin rate of the generator rotor.

Answer C is incorrect. Only Technician A is correct.

Answer D is incorrect. Technician A is correct.

TASK C.5

26. An inoperative generator is being diagnosed. After full-fielding the generator, it begins to charge at full capacity. Technician A says that the generator mounting bracket could be loose. Technician B says that a faulty voltage regulator could be the cause. Who is correct?

A. A only
B. B only
C. Both A and B
D. Neither A nor B

Answer A is incorrect. A loose generator mounting bracket could cause a voltage drop problem in the charging ground circuit. Full-fielding the generator would not mask this problem.

Answer B is correct. Only Technician B is correct. Full-fielding the generator bypasses the voltage regulator and should only be done for a few seconds. If the generator begins to charge while full-fielding, the problem is the voltage regulator.

Answer C is incorrect. Only Technician B is correct.

Answer D is incorrect. Technician B is correct.

TASK C.6

27. What is the most likely test that could prove that a poor connection was present at the charging insulated circuit?

A. Voltage drop test of the charging ground circuit
B. Voltage drop test of the negative battery cable
C. Voltage drop test of the positive battery cable
D. Voltage drop test of the charging output wire

Answer A is incorrect. The charging ground is not considered the insulated side of the charging circuit.

Answer B is incorrect. Testing the negative battery cable is most effective when done while cranking the engine.

Answer C is incorrect. Testing the positive battery cable is most effective when done while cranking the engine.

Answer D is correct. Performing a voltage drop test in the charging output wire will test the quality of the wire and associated connections. This test should be done while the engine is running with some electrical loads turned on.

28. A vehicle has a problem of the headlights on the right side not being as bright as the headlights on the left side. Technician A says that a bad dimmer switch could be the cause. Technician B says that a loose ground connection on the right side could be the cause. Who is correct?

TASK D.1

A. A only

B. B only

C. Both A and B

D. Neither A nor B

Answer A is incorrect. The dimmer switch supplies power to both headlights and would not likely cause one side to be less bright.

Answer B is correct. Only Technician B is correct. A poor ground on the right side headlight could cause that side to be less bright due to the likely voltage drop at that location.

Answer C is incorrect. Only Technician B is correct.

Answer D is incorrect. Technician B is correct.

29. Technician A says that the daytime running lights utilize the turn signal filaments on most late model vehicles. Technician B says that the daytime running lights will use more current than the headlights. Who is correct?

TASK D.2

A. A only

B. B only

C. Both A and B

D. Neither A nor B

Answer A is incorrect. The turn signal filaments are not used as daytime running lights on most late model vehicles. The headlight filaments or the running lights are most often used as daytime running lights.

Answer B is incorrect. The daytime running lights would not normally use more current than the headlights due to the typical lower intensity of these lights.

Answer C is incorrect. Neither Technician is correct.

Answer D is correct. Neither Technician is correct. The headlights or running lights are the typical bulbs used for daytime running lights. The daytime running lights typically are not as bright as the headlights and would not likely require more current.

TASK D.3

30. The fog lights are dimmer than normal and a voltage drop test across the bulbs reveals 8 volts. Which of the following conditions would most likely cause this problem?

A. Faulty fog light bulbs
B. Faulty fog light switch contacts
C. Faulty fog light fuse
D. Faulty turn signal switch

Answer A is incorrect. There is no indication of faulty bulbs. The voltage test proved that only 8 volts is being delivered to the bulbs.

Answer B is correct. The fog lights should receive about 12 to 13 volts when they are turned on. Since the test revealed only 8 volts, there has to be a voltage loss at some point. The fog light switch could be a point of voltage loss.

Answer C is incorrect. A bad fuse would not typically cause low voltage to be delivered to fog light bulbs. If the fuse were bad, there would be zero volts at the lights and they would be inoperative.

Answer D is incorrect. A turn signal switch would not cause the low voltage to be present at the fog lights.

TASK D.5

31. All of the following conditions could cause the stop lights to be inoperative EXCEPT:

A. A closed stop light switch
B. Open stop light switch
C. Blown stop light fuse
D. Open turn signal switch

Answer A is correct. A closed stop light switch would cause the lights to operate all of the time.

Answer B is incorrect. An open stop light switch could cause the stop lights to be inoperative.

Answer C is incorrect. A blown stop light fuse would cause the stop lights to be inoperative.

Answer D is incorrect. An open turn signal switch could cause the stop lights to be inoperative. The stop light circuit routes through the turn signal switch on many late model vehicles that use the same filament for the turn and stop lights.

TASK D.5

32. Technician A says that turn signal flashers are sensitive to electrical load. Technician B says that hazard flashers are not sensitive to electrical load. Who is correct?

A. A only
B. B only
C. Both A and B
D. Neither A nor B

Answer A is incorrect. Technician B is also correct.

Answer B is incorrect. Technician A is also correct.

Answer C is correct. Both Technicians are correct. Turn signal flashers will change the flash rate when the electrical load changes. This characteristic provides feedback to the driver when a bulb blows or extra bulbs are connected. The hazard flashers will not typically change the flash rate when the current rate changes.

Answer D is incorrect. Both Technicians are correct.

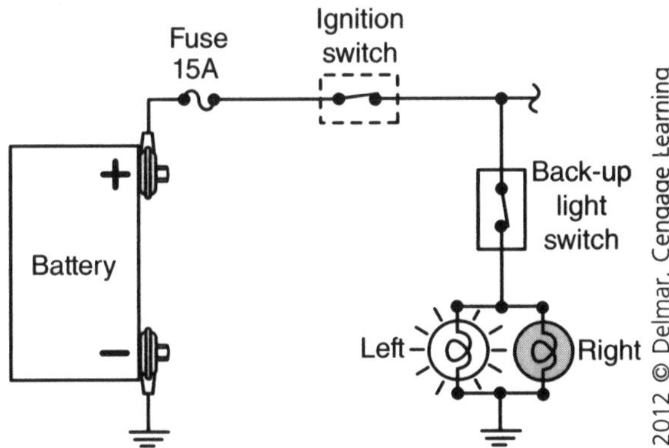

33. Referring to the figure, the back-up lamp on the right side is inoperative but the left side works fine. Technician A says that the right side lamp socket could be open. Technician B says that the right side bulb could be open. Who is correct?

TASK E.2.5

A. A only

B. B only

C. Both A and B

D. Neither A nor B

Answer A is incorrect. Technician B is also correct.

Answer B is incorrect. Technician A is also correct.

Answer C is correct. Both Technicians are correct. An open bulb or socket on the right side could cause the right side lamp to be inoperative. The left side would continue functioning with either of the problems described in the question.

Answer D is incorrect. Both Technicians are correct.

34. A vehicle is being diagnosed that has a fuel gauge that does not work. Technician A says that the other gauges should be checked for correct operation as part of the diagnosis process. Technician B says that the sending unit for the fuel gauge is in the fuel tank. Who is correct?

TASK E.3

A. A only

B. B only

C. Both A and B

D. Neither A nor B

Answer A is incorrect. Technician B is also correct.

Answer B is incorrect. Technician A is also correct.

Answer C is correct. Both Technicians are correct. It is a good diagnostic technique to verify that the other gauges are functioning correctly. The sending unit for the fuel gauge is located in the fuel tank.

Answer D is incorrect. Both Technicians are correct.

TASK E.5

35. What type of sending unit does the temperature gauge use?

A. Rheostat

B. Potentiometer

C. Photo resistor

D. Thermistor

Answer A is incorrect. A rheostat is typically used as the sending unit for the fuel gauge.

Answer B is incorrect. A potentiometer is sometimes used as the sending unit for the fuel gauge.

Answer C is incorrect. A photo resistor is used as an input to automatic lighting systems as well as some automatic temperature control HVAC systems.

Answer D is correct. A thermistor is typically used as the sending unit for the temperature gauge.

TASKS E.3, E.5

36. A vehicle is being diagnosed for a problem of all of the gauges being inoperative. The vehicle is equipped with an electronic cluster that has self-diagnostic capabilities. Technician A says that a scan tool should be connected to the data link connector to retrieve cluster data. Technician B says that this problem could be diagnosed by using the self-diagnostic function of the instrument cluster. Who is correct?

A. A only

B. B only

C. Both A and B

D. Neither A nor B

Answer A is incorrect. Technician B is also correct.

Answer B is incorrect. Technician A is also correct.

Answer C is correct. Both Technicians are correct. Many late model instrument clusters have functioning processors (computers). These instrument panels often have self-diagnostic capabilities that can be accessed by pressing the buttons in a certain sequence. These panels also can be diagnosed with scan tools.

Answer D is incorrect. Both Technicians are correct.

TASK E.5

37. All of the following types of electronic devices are used as inputs to electronic gauge assemblies EXCEPT:

A. Thermistor

B. Piezo resistor

C. Diode

D. Rheostat

Answer A is incorrect. Thermistors are used as inputs to electronic temperature gauges.

Answer B is incorrect. Piezo resistors are used as inputs to electronic oil pressure gauges.

Answer C is correct. Diodes are not used as inputs to electronic gauges assemblies. Diodes are used widely inside computers as well as clamping devices for electromagnetic coils.

Answer D is incorrect. Rheostats are used as inputs to many electronic fuel gauges.

38. The temperature light does not illuminate during the bulb test or when the vehicle overheats. The most likely cause for this condition would be?

TASK E.4

A. Temperature sending unit
B. Bulb is open
C. Grounded wire to the sending unit
D. Open wire to the sending unit

Answer A is incorrect. A temperature sending unit would not cause the temperature light to be inoperative during the bulb test.

Answer B is correct. An open temperature light bulb would cause the temperature light to be inoperative at all times.

Answer C is incorrect. A grounded wire to the temperature sending unit would cause the temperature light to illuminate continuously.

Answer D is incorrect. An open wire to the sending unit would not cause the temperature light to be inoperative during the bulb test.

39. All of the following items affect the driver information center EXCEPT:

TASK E.5

A. Oxygen sensor
B. Power train control module
C. Body control module
D. Data bus network

Answer A is correct. The oxygen sensor typically connects to the power train control module and provides data about the oxygen content in the exhaust system.

Answer B is incorrect. The power train control module (PCM) does provide data to the driver information center over the data bus network.

Answer C is incorrect. The body control module (BCM) does provide data to the driver information center over the data bus network.

Answer D is incorrect. The data bus network is connected to the driver information center and is an "information highway" that allows many computers to send and receive data messages.

40. Technician A says that the horn relay can be activated by the BCM. Technician B says that the horn relay can be activated by a ground signal that is received from the horn switch. Who is correct?

TASK F.9

A. A only
B. B only
C. Both A and B
D. Neither A nor B

Answer A is incorrect. Technician B is also correct.

Answer B is incorrect. Technician A is also correct.

Answer C is correct. Both Technicians are correct. The horn relay typically operates by receiving a ground at the relay coil. Some manufacturers use the BCM to make this happen at the correct time while other manufacturers connect the circuit directly to the horn switch.

Answer D is incorrect. Both Technicians are correct.

TASK F.10

41. The horn does not operate when the horn switch is depressed but it will operate when the panic button is depressed on the keyless entry transmitter. Technician A says that the relay coil could be shorted. Technician B says that the horn fuse could be blown. Who is correct?

A. A only
B. B only
C. Both A and B
D. Neither A nor B

Answer A is incorrect. A shorted relay coil would cause the horn to be totally inoperative at all times.

Answer B is incorrect. A blown horn fuse would cause the horn to be totally inoperative at all times.

Answer C is incorrect. Neither Technician is correct.

Answer D is correct. Neither Technician is correct. Since the horn operates while pressing the panic button, neither of the problems listed in the question could be the cause. A faulty horn switch or a faulty clock spring could be the cause of the problem.

TASK F.9

42. What is the most likely problem that could cause the windshield wipers to continue running after the wiper switch is turned off?

A. Open wire between the wiper switch and the wiper motor
B. Shorted wiper wash motor switch
C. Shorted wiper park switch
D. Open wiper relay

Answer A is incorrect. An open wire between the switch and the wiper motor would likely cause the wipers to be inoperative in one of the settings.

Answer B is incorrect. A shorted wiper wash motor switch would likely cause the wash motor to exhibit a problem.

Answer C is correct. A shorted wiper park switch can cause the wipers to run continuously.

Answer D is incorrect. An open wiper relay would cause the wipers to be inoperative in one of the functions.

TASK F.1

43. All of the power windows operate slowly in both the up and down positions. Technician A says that the ground for the master switch could have a loose connection. Technician B says that a wrong fuse could be installed for the circuit. Who is correct?

A. A only
B. B only
C. Both A and B
D. Neither A nor B

Answer A is correct. Only Technician A is correct. A loose ground connection could cause the power windows to operate slowly due to the likely voltage drop at that location.

Answer B is incorrect. An incorrect fuse would not cause slow operation of the power windows. If the fuse were too small, then it would blow and cause the power windows to be totally inoperative.

Answer C is incorrect. Only Technician A is correct.

Answer D is incorrect. Technician A is correct.

44. A "scissor style" power window regulator is being replaced on a vehicle. Technician A says this can be completed without removing the motor from the door skin. Technician B says care should be taken when disconnecting the motor from the regulator due to the spring tension on the regulator. Who is correct?

TASK F.2

A. A only
B. B only
C. Both A and B
D. Neither A nor B

Answer A is incorrect. The power window motor and regulator would have to be removed from the vehicle to replace the regulator.

Answer B is correct. Only Technician B is correct. This style of power window system uses a strong coil spring that can be dangerous during the service of the motor or regulator. The regulator should be secured prior to disconnecting the motor.

Answer C is incorrect. Only Technician B is correct.

Answer D is incorrect. Technician B is correct.

45. Technician A says that the rear window defogger uses heated air to defog the back glass. Technician B says that the rear window defogger operates on a timer in order to limit the heavy electrical load on the vehicle. Who is correct?

TASK F.3

A. A only
B. B only
C. Both A and B
D. Neither A nor B

Answer A is incorrect. Most rear defoggers use a heat strip located on the rear glass to defog the area.

Answer B is correct. Only Technician B is correct. Rear defoggers use a timer to limit the heavy electrical load on the vehicle. The timer is usually set to turn the system off after approximately 15 minutes.

Answer C is incorrect. Only Technician B is correct.

Answer D is incorrect. Technician B is correct.

46. The keyless entry system does not work with remote transmitter #1 but works correctly with remote transmitter #2. Technician A says that the battery in transmitter #1 may be weak and need to be replaced. Technician B says that transmitter #1 could be defective and need to be replaced. Who is correct?

TASK F.5

A. A only
B. B only
C. Both A and B
D. Neither A nor B

Answer A is incorrect. Technician B is also correct.

Answer B is incorrect. Technician A is also correct.

Answer C is correct. Both Technicians are correct. A weak battery or a defective transmitter could cause the problem in this question.

Answer D is incorrect. Both Technicians are correct.

TASK F.2

47. The radio has poor reception and has a buzzing sound that increases with engine speed. The operation of the sound system is normal when a CD is used in the player. Technician A says that the main sound system ground could have a loose connection. Technician B says that the antenna coaxial cable could be damaged. Who is correct?

A. A only
B. B only
C. Both A and B
D. Neither A nor B

Answer A is incorrect. A main ground problem would also cause a problem for the CD player.

Answer B is correct. Only Technician B is correct. A damaged antenna coaxial cable could cause the radio to have poor operation but it would not affect the CD player.

Answer C is incorrect. Only Technician B is correct.

Answer D is incorrect. Technician B is correct.

TASK F.2

48. The auxiliary power outlet is inoperative and the fuse is found to be open. What is the most likely cause for this condition?

A. Loose connection at the power outlet plug
B. Foreign metal object in the power outlet
C. Broken wire leading to the power outlet
D. Open internal connection at the power outlet

Answer A is incorrect. A loose connection would not typically blow the fuse. This would just cause the components that connect to the outlet to not operate correctly.

Answer B is correct. It is not uncommon for metal objects to fall into the auxiliary power outlets, which can cause a short to ground and overload the fuse.

Answer C is incorrect. A broken wire does not typically cause a fuse to blow.

Answer D is incorrect. An open connection at the power outlet would not cause the fuse to blow.

TASK F.9

49. All of the following conditions could cause the airbag light to stay on while the vehicle is in operation EXCEPT:

A. An open clock spring
B. A shorted airbag inflator
C. A faulty seatbelt buckle
D. An open airbag light bulb

Answer A is incorrect. An open clock spring could cause the airbag light to stay on while driving. The airbag controller is continuously monitoring the airbag circuits for problems.

Answer B is incorrect. A shorted airbag inflator would cause the airbag light to stay on while driving. The airbag controller is continuously monitoring the airbag circuits for problems.

Answer C is incorrect. A faulty seatbelt buckle could cause the airbag light to stay on while driving. The airbag controller is continuously monitoring the airbag circuits for problems.

Answer D is correct. An open airbag light bulb would cause the light to be totally inoperative.

50. An airbag system needs to be disarmed during a dash pad replacement. Technician A says that it is necessary to remove power from the system prior to component removal. Technician B says that the inflatable devices should be laid "face up" in a secure area after being removed from a vehicle. Who is correct?

A. A only

B. B only

C. Both A and B

D. Neither A nor B

Answer A is incorrect. Technician B is also correct.

Answer B is incorrect. Technician A is also correct.

Answer C is correct. Both Technicians are correct. It is advisable to remove the power supply prior to airbag component removal. The best way to do this is to remove the negative battery cable. Live inflatable devices should be stored "face up" in a secure area to prevent the device from being accidentally deployed and then projected up in the air.

Answer D is incorrect. Both Technicians are correct.

PREPARATION EXAM 2—ANSWER KEY

1. C
2. B
3. C
4. A
5. C
6. C
7. A
8. D
9. C
10. A
11. C
12. B
13. C
14. C
15. B
16. C
17. A
18. A
19. C
20. B
21. C
22. A
23. D
24. A
25. D
26. C
27. B
28. C
29. B
30. B
31. A
32. C
33. C
34. D
35. A
36. D
37. D
38. C
39. C
40. D
41. B
42. D
43. C
44. A
45. B
46. C
47. D
48. A
49. C
50. B

PREPARATION EXAM 2—EXPLANATIONS

TASK F.3

1. A late-model luxury vehicle is being diagnosed for a problem in the driver's side heated seat. The seat stays hot at all times regardless of the seat heater switch position. Technician A says that a shorted seat heater relay could cause this problem. Technician B says that a grounded circuit between the seat heater relay and the body control module (BCM) could be the cause. Who is correct?

   A. A only
   B. B only
   C. Both A and B
   D. Neither A nor B

   Answer A is incorrect. Technician B is also correct.

   Answer B is incorrect. Technician A is also correct.

   Answer C is correct. Both Technicians are correct. A shorted seat heater relay would cause the seat heater to receive current at all times. A grounded circuit between the seat heater relay and the BCM would cause the relay to be activated at all times. Both of these problems would cause the seat to stay hot at all times.

   Answer D is incorrect. Both Technicians are correct.

2. All of the following statements about performing voltage drop tests on electrical circuits are true EXCEPT:

TASK A.1

   A. The voltmeter leads should be connected on each end of the circuit to be tested.
   B. The meter should be connected in series with the circuit.
   C. The circuit should be energized when performing a voltage drop test.
   D. The digital meter should be set to the DC volts scale.

   Answer A is incorrect. The voltmeter leads need to be connected on each end of the circuit that is being tested while the circuit is energized.

   Answer B is correct. The voltmeter must be connected in parallel with the electrical circuit that is being tested.

   Answer C is incorrect. The voltmeter leads need to be connected on each end of the circuit that is being tested while the circuit is energized.

   Answer D is incorrect. The voltmeter should be connected on the DC volts scale to test for voltage drops on most automotive circuits.

3. Technician A says that a poor ground connection will cause reduced current flow in an electrical circuit. Technician B says that water corrosion in the wiring will cause reduced current flow in an electrical circuit. Who is correct?

TASK A.2

   A. A only
   B. B only
   C. Both A and B
   D. Neither A nor B

   Answer A is incorrect. Technician B is also correct.

   Answer B is incorrect. Technician A is also correct.

   Answer C is correct. Both Technicians are correct. A poor ground and corroded wiring will both decrease current flow in an electrical circuit due to the increased resistance associated with both of these problems.

   Answer D is incorrect. Both Technicians are correct.

4. A resistance test was performed on a closed brake switch and the result is .857 megohms. Technician A says that the switch is faulty because the reading is beyond the specifications. Technician B says that the switch has 857 ohms of resistance. Who is correct?

TASK A.3

   A. A only
   B. B only
   C. Both A and B
   D. Neither A nor B

   Answer A is correct. Only Technician A is correct. A closed switch should have electrical continuity, which would be a very low resistance value.

   Answer B is incorrect. The reading would be 857,000 ohms.

   Answer C is incorrect. Only Technician A is correct.

   Answer D is incorrect. Technician A is correct.

TASK A.4

5. Technician A says that an oscilloscope shows voltage signals represented as lines on a digital screen. Technician B says that an oscilloscope can be used to view live sensor readings directly at each sensor. Who is correct?

   A. A only
   B. B only
   C. Both A and B
   D. Neither A nor B

Answer A is incorrect. Technician B is also correct.

Answer B is incorrect. Technician A is also correct.

Answer C is correct. Both Technicians are correct. Oscilloscopes are tools that a technician can use to view electrical signals and voltages. These readings show up on a digital screen and can even be taken directly at each sensor.

Answer D is incorrect. Both Technicians are correct.

TASK A.5

6. Technician A says that a scan tool can be used to retrieve a trouble code from the body control module. Technician B says that a scan tool can be used to view live sensor data from the body control module. Who is correct?

   A. A only
   B. B only
   C. Both A and B
   D. Neither A nor B

Answer A is incorrect. Technician B is also correct.

Answer B is incorrect. Technician A is also correct.

Answer C is correct. Both Technicians are correct. Scan tools can be used to view live data or to retrieve trouble codes from electronic modules on late model vehicles.

Answer D is incorrect. Both Technicians are correct.

TASK F.7

7. Voice activation commands are used on some luxury vehicles to interact with all of the following accessory systems EXCEPT:

   A. Power windows
   B. Audio system
   C. Navigation system
   D. Mobile phone system

Answer A is correct. Voice activation systems typically do not have an input to the power windows.

Answer B is incorrect. Vehicles with voice activation allow oral commands to operate the audio system.

Answer C is incorrect. Vehicles with voice activation allow oral commands to operate the navigation system.

Answer D is incorrect. Vehicles with voice activation allow oral commands to operate the mobile phone system.

8. Technician A says that a broken wire will often cause a fuse to blow due to high current flow. Technician B says that a corroded wire connection will often cause a fuse to blow due to high current flow. Who is correct?

TASK A.6

A. A only

B. B only

C. Both A and B

D. Neither A nor B

Answer A is incorrect. A broken wire does not allow current to flow so this problem would likely never cause a fuse to blow.

Answer B is incorrect. Corrosion that invades electrical wiring and connections adds electrical resistance to the circuit, which would reduce the current flow.

Answer C is incorrect. Neither Technician is correct.

Answer D is correct. Neither Technician is correct. Broken wires and corrosion would not typically cause a fuse to blow. Failures that increase current flow such as grounded circuits, physical resistance and short circuits are some typical causes for blown fuses.

9. A vehicle is in the repair shop for a dead battery. During the diagnosis, an excessive parasitic draw is found. Technician A says that removing one fuse at a time is an effective method of locating the unwanted draw. Technician B says that a "stuck closed" trunk light switch could be the cause. Who is correct?

TASK A.7

A. A only

B. B only

C. Both A and B

D. Neither A nor B

Answer A is incorrect. Technician B is also correct.

Answer B is incorrect. Technician A is also correct.

Answer C is correct. Both Technicians are correct. A technician should always look for circuits that are staying on when a high parasitic draw is experienced. If no apparent circuit is found, then pulling the fuses is a way to isolate which circuit is causing the draw.

Answer D is incorrect. Both Technicians are correct.

10. Technician A says that tugging on a fusible link is an acceptable method of quickly testing it. Technician B says that a fusible link can be tested with an ohmmeter without disconnecting it from the vehicle. Who is correct?

TASK A.8

A. A only

B. B only

C. Both A and B

D. Neither A nor B

Answer A is correct. Only Technician A is correct. Fusible links can be quickly tested by pulling on the assembly. If it is not damaged, it will not be affected. If the fusible link is blown, then it will separate when pulling on it.

Answer B is incorrect. A circuit breaker will need to be disconnected from the vehicle if it is going to be tested with an ohmmeter.

Answer C is incorrect. Only Technician A is correct.

Answer D is incorrect. Technician A is correct.

TASKS
A.8, F.1

11. Which of the following conditions would LEAST LIKELY cause a fusible link for the power windows to burn up?

A. A motor with a shorted winding

B. A power-side short to ground

C. High resistance in the switch contacts

D. A binding window regulator

Answer A is incorrect. A shorted motor winding would cause a rapid increase in electrical current flow, which would cause the fusible link to burn up.

Answer B is incorrect. A power-side short to ground would cause high levels of current flow, which would cause the fusible link to burn up.

Answer C is correct. High resistance would not increase current flow. This problem would reduce current flow and the fusible link would not likely burn up.

Answer D is incorrect. A binding window regulator would cause an increase in current flow, which could cause the fusible link to burn up.

TASK A.9

12. Which of the following details would be the LEAST LIKELY item to be located on a wiring diagram?

A. The circuit number of a wire

B. A flowchart for troubleshooting an electrical problem

C. The amp rating of the fuse that supplies the circuit

D. The color of a wire

Answer A is incorrect. Wiring diagrams will typically display the circuit number for all of the wires shown.

Answer B is correct. Wiring diagrams do not typically contain any type of flow chart. However, an experienced technician can use a wiring diagram to effectively troubleshoot problems in electrical circuits.

Answer C is incorrect. Wiring diagrams will usually show the amp rating any fuses that are shown.

Answer D is incorrect. Wiring diagrams will typically show the wire color for all of the wires shown.

TASK A.10

13. Technician A states that the scan tool receives data from the data bus network. Technician B states that a bi-directional scan tool can send functional messages over the data bus network. Who is correct?

A. A only

B. B only

C. Both A and B

D. Neither A nor B

Answer A is incorrect. Technician B is also correct.

Answer B is incorrect. Technician A is also correct.

Answer C is correct. Both Technicians are correct. Scan tools can be used to connect to the data bus network. A scan tool can read data and trouble codes from a vehicle as well as send output test commands to a vehicle. These functional tests are valuable to the technician during the diagnosis of electrical problems.

Answer D is incorrect. Both Technicians are correct.

14. A vehicle is in the repair shop for a battery problem. Technician A says that the battery voltage should be 12.4 volts before performing a battery load test. Technician B says that a battery can be accurately tested with a digital tester if it has at least 12 volts. Who is correct?

TASKS B.1, B.2

A. A only
B. B only
C. Both A and B
D. Neither A nor B

Answer A is incorrect. Technician B is also correct.

Answer B is incorrect. Technician A is also correct.

Answer C is correct. Both Technicians are correct. A battery must have at least 12.4 volts before performing a valid load test. 12.4 volts represents a 75 percent charge level. A battery can be accurately tested with a digital battery tester if it has at least 12 volts. These testers perform a capacitance test on the battery and will let the technician know if the battery needs to be charged up or replaced.

Answer D is incorrect. Both Technicians are correct.

15. A battery is being replaced in a vehicle. Technician A says that the replacement battery should have at least 75 percent of the original battery CCA rating for the vehicle. Technician B says that the replacement battery should not exceed the physical dimensions of the original battery. Who is correct?

TASK B.5

A. A only
B. B only
C. Both A and B
D. Neither A nor B

Answer A is incorrect. All replacement batteries should meet the original specifications of the original battery.

Answer B is correct. Only Technician B is correct. The replacement must physically fit into the limited space that is usually provided for the vehicle battery.

Answer C is incorrect. Only Technician B is correct.

Answer D is incorrect. Technician B is correct.

16. Technician A says that batteries can be recharged more quickly by using a high setting on the battery charger. Technician B says that batteries can be more thoroughly charged by using a low setting on the battery charger. Who is correct?

TASK B.4

A. A only
B. B only
C. Both A and B
D. Neither A nor B

Answer A is incorrect. Technician B is also correct.

Answer B is incorrect. Technician A is also correct.

Answer C is correct. Both Technicians are correct. Batteries can be charged on a high setting more quickly. However, the technician should monitor the voltage and temperature of the battery in order to not overcharge the battery. A slow charge typically provides a more thorough charge but it takes more time to do.

Answer D is incorrect. Both Technicians are correct.

TASK B.6

17. A vehicle needs to be jumpstarted with an auxiliary power supply. Technician A says that safety glasses should always be worn when working around batteries. Technician B says that the positive connection should be made at the starter solenoid. Who is correct?

A. A only
B. B only
C. Both A and B
D. Neither A nor B

Answer A is correct. Only Technician A is correct. Safety glasses should always be worn when working around vehicle batteries.

Answer B is incorrect. The positive connection should be made at the positive battery terminal. It is recommended to connect the ground connection to a quality engine ground to prevent a potential spark at the battery.

Answer C is incorrect. Only Technician A is correct.

Answer D is incorrect. Technician A is correct.

TASK B.7

18. Technician A says that a shorted armature can cause elevated starter current draw. Technician B says that a corroded battery cable terminal can cause elevated starter current draw. Who is correct?

A. A only
B. B only
C. Both A and B
D. Neither A nor B

Answer A is correct. Only Technician A is correct. A shorted winding or a short to ground will both cause increased current flow in the circuit.

Answer B is incorrect. Since corrosion creates resistance to current flow, corroded cables would reduce current flow to the starter.

Answer C is incorrect. Only Technician A is correct.

Answer D is incorrect. Technician A is correct.

TASK B.7

19. A vehicle with a six cylinder engine is being diagnosed for a starting problem. A starter current draw test was performed and the amperage was 310 amps. Technician A says that this reading could be caused by a weak battery. Technician B says that this reading could be caused by tight starter bushings. Who is correct?

A. A only
B. B only
C. Both A and B
D. Neither A nor B

Answer A is incorrect. Technician B is also correct.

Answer B is incorrect. Technician A is also correct.

Answer C is correct. Both Technicians are correct. 310 amps of starter draw on a six cylinder engine is above normal. A weak battery will cause the starter draw to rise above normal. Tight starter bushings will also cause the starter current draw to be higher than normal.

Answer D is incorrect. Both Technicians are correct.

20. All of the following components are parts of the starter control circuit EXCEPT:

TASK B.9

A. Park/neutral switch
B. Positive battery cable
C. Starter relay
D. Ignition switch

Answer A is incorrect. The park/neutral switch is in the starter control circuit to prevent the starter from operating when the vehicle is in gear.

Answer B is correct. The positive battery cable is in the starter (high current/load) circuit.

Answer C is incorrect. The starter relay is in the starter control circuit to limit the amount of current flowing through the ignition switch.

Answer D is incorrect. The ignition switch is in the starter control circuit to allow the driver to initiate the start sequence.

21. Technician A says that the starter solenoid provides the linear movement to push the starter drive gear into the flywheel. Technician B says that the starter solenoid acts as a magnetic switch to provide a current path for the positive circuit to reach the starter motor. Who is correct?

TASK B.9

A. A only
B. B only
C. Both A and B
D. Neither A nor B

Answer A is incorrect. Technician B is also correct.

Answer B is incorrect. Technician A is also correct.

Answer C is correct. Both Technicians are correct. The starter solenoid provides linear movement to engage the drive gear. The starter solenoid also works as a magnetic switch to provide a current path for the positive circuit to reach the starter motor.

Answer D is incorrect. Both Technicians are correct.

22. Technician A says that the negative battery cable should be removed prior to disconnecting the electrical connections at the starter. Technician B says that the fasteners should be removed prior to removing the electrical connections at the starter. Who is correct?

TASK B.9

A. A only
B. B only
C. Both A and B
D. Neither A nor B

Answer A is correct. Only Technician A is correct. It is advisable to always disconnect the battery prior to removing the starter motor. It is safer to remove the negative cable first due to the possibility of letting the wrench touch metal while connected to the cable end.

Answer B is incorrect. The electrical connections should be removed before removing the fasteners, if possible, so that the starter does not have to hang by the connecting wires.

Answer C is incorrect. Only Technician A is correct.

Answer D is incorrect. Technician A is correct.

TASK C.1

23. Which of the following conditions would be the LEAST LIKELY cause of an undercharged battery?

    A. Worn generator brushes
    B. Loose fastener at the charge wire connection
    C. Faulty voltage regulator
    D. Battery cable that is oversized

Answer A is incorrect. Worn generator brushes would cause low charging system output resulting in an undercharged battery.

Answer B is incorrect. A loose fastener at the charge wire connection would reduce charging output resulting in an undercharged battery.

Answer C is incorrect. A voltage regulator that reduces charging output could cause an undercharged battery.

Answer D is correct. Having a larger battery cable would not cause an undercharged battery. The larger cable would have less resistance than a smaller one.

TASK C.2

24. The accessory drive belt system should be inspected during regular intervals. Technician A says that a serpentine drive belt tensioner should snap back after releasing pressure on it. Technician B says that the drive belt should be replaced at the first sign of cracks on the back side of it. Who is correct?

    A. A only
    B. B only
    C. Both A and B
    D. Neither A nor B

Answer A is correct. Only Technician A is correct. A drive belt tensioner should have good spring tension when released.

Answer B is incorrect. Small cracks found during a belt inspection should be noted on the repair order. Large cracks and/or missing pieces from the belt are a cause for the belt to be replaced.

Answer C is incorrect. Only Technician A is correct.

Answer D is incorrect. Technician A is correct.

TASK C.4

25. What is the LEAST LIKELY cause for low charging current during a charging output test?

    A. Worn brushes
    B. Faulty rectifier bridge
    C. Shorted stator winding
    D. Faulty charging gauge

Answer A is incorrect. Worn generator brushes will cause low charging output.

Answer B is incorrect. A faulty rectifier bridge will cause low charging output.

Answer C is incorrect. Shorted stator windings will cause low charging output.

Answer D is correct. A faulty charging gauge is not a typical cause for low charging output during a charging output test.

26. A vehicle with a charging problem is being repaired. The charging output wire received damage and needed to be repaired. Technician A says to use weather resistant connectors when making wire repairs in the engine compartment. Technician B says that solder and heat shrink is an acceptable method of wire repair in the engine compartment. Who is correct?

TASK C.7

A. A only
B. B only
C. Both A and B
D. Neither A nor B

Answer A is incorrect. Technician B is also correct.

Answer B is incorrect. Technician A is also correct.

Answer C is correct. Both Technicians are correct. Water resistant wire repair methods should always be used when making repairs in the engine compartment. Weather resistant connectors as well as solder and heat shrink are both acceptable ways to perform wire repairs in the engine compartment.

Answer D is incorrect. Both Technicians are correct.

27. A generator is being replaced. Technician A uses an air tool to remove the fastening nut from the charging output wire. Technician B disconnects the negative battery cable prior to removing the generator. Who is correct?

TASK C.8

A. A only
B. B only
C. Both A and B
D. Neither A nor B

Answer A is incorrect. An air tool should never be used to remove the fastening nut from the charging output wire. Doing this will likely damage the internal connections of the generator.

Answer B is correct. Only Technician B is correct. It is a good practice to remove the negative battery cable prior to removing the generator.

Answer C is incorrect. Only Technician B is correct.

Answer D is incorrect. Technician B is correct.

28. Technician A uses an alignment tool to aim the headlights on a late model vehicle. Technician B uses the built-in alignment bubbles on the headlamp housings to aim the headlights on some late model vehicles. Who is correct?

TASK D.2

A. A only
B. B only
C. Both A and B
D. Neither A nor B

Answer A is incorrect. Technician B is also correct.

Answer B is incorrect. Technician A is also correct.

Answer C is correct. Both Technicians are correct. The headlights can be adjusted on some vehicles by using an alignment tool to provide feedback on the height and direction of the headlights. Some vehicles have built-in alignment bubbles to assist the technician with aligning the headlights.

Answer D is incorrect. Both Technicians are correct.

TASK D.1

29. The left rear park light bulb is inoperative. The turn signal and stop lamp bulbs work normally on that side. Technician A says that a faulty ground on the left rear could be the cause. Technician B says that a faulty bulb could be the cause. Who is correct?

A. A only
B. B only
C. Both A and B
D. Neither A nor B

Answer A is incorrect. A faulty left rear lighting ground would likely affect the turn and stop lamps on that side of the vehicle.

Answer B is correct. Only Technician B is correct. A faulty park light bulb could cause the problem described.

Answer C is incorrect. Only Technician B is correct.

Answer D is incorrect. Technician B is correct.

TASK E.2

30. A vehicle electronic instrument cluster lighting intermittently goes dark after being driven for long periods. Technician A says the problem could be a shorted LED. Technician B says the problem could be a faulty lighting driver in the instrument cluster assembly. Who is correct?

A. A only
B. B only
C. Both A and B
D. Neither A nor B

Answer A is incorrect. A shorted LED would not work at all.

Answer B is correct. Only Technician B is correct. A faulty lighting driver could cause the intermittent problem. Some electronic devices are sensitive to heat and being used for long periods of time.

Answer C is incorrect. Only Technician B is correct.

Answer D is incorrect. Technician B is correct.

TASKS D.3, D.5

31. A vehicle is being diagnosed for inoperative brake lights. Technician A says that a blown stop light fuse could be the problem. Technician B says that a shorted headlight switch could be the cause. Who is correct?

A. A only
B. B only
C. Both A and B
D. Neither A nor B

Answer A is correct. Only Technician A is correct. A blown stop light fuse could cause the brake lights to be inoperative.

Answer B is incorrect. A shorted headlight switch would not likely cause any problem for the stop lights.

Answer C is incorrect. Only Technician A is correct.

Answer D is incorrect. Technician A is correct.

32. A vehicle is being diagnosed for inoperative turn signals. The hazard lights work correctly. Technician A says that a faulty turn signal flasher could be the cause. Technician B says that a faulty multi-function switch could be the cause. Who is correct?

TASK D.4

A. A only

B. B only

C. Both A and B

D. Neither A nor B

Answer A is incorrect. Technician B is also correct.

Answer B is incorrect. Technician A is also correct.

Answer C is correct. Both Technicians are correct. A faulty turn signal flasher could cause the turn signals to be inoperative. The turn signal switch is combined into the multi-function switch on many late-model vehicles.

Answer D is incorrect. Both Technicians are correct.

33. Technician A says that the hazard flasher on late model vehicles can be replaced without disconnecting the vehicle battery. Technician B says that the turn signal flasher and hazard flasher are combined into one assembly on some late model vehicles. Who is correct?

TASK D.5

A. A only

B. B only

C. Both A and B

D. Neither A nor B

Answer A is incorrect. Technician B is also correct.

Answer B is incorrect. Technician A is also correct.

Answer C is correct. Both Technicians are correct. The vehicle battery would not have to be disconnected to replace the turn signal flasher assembly. Some vehicles have combined the turn flasher and the hazard flasher into one assembly.

Answer D is incorrect. Both Technicians are correct.

34. The back-up light switch is being tested on a vehicle. Technician A says that the back-up light switch is often combined with the turn signal switch. Technician B says that the back-up light switch is located near the rear light housing. Who is correct?

TASK D.5

A. A only

B. B only

C. Both A and B

D. Neither A nor B

Answer A is incorrect. The back-up light switch is not typically combined into the turn signal switch. The back-up light switch is typically mounted on or near the transmission/transaxle.

Answer B is incorrect. The back-up light switch is not located near the rear light housing. The back-up light switch is typically mounted on or near the transmission/transaxle.

Answer C is incorrect. Neither Technician is correct.

Answer D is correct. Neither Technician is correct. The back-up light switch is typically mounted on or near the transmission/transaxle.

TASK E.4

35. A vehicle is being diagnosed with a problem of the door ajar alarm sounding intermittently. Technician A says that the striker adjustment of all doors should be checked. Technician B says that the door jam switches should be replaced. Who is correct?

A. A only
B. B only
C. Both A and B
D. Neither A nor B

Answer A is correct. Only Technician A is correct. A misadjusted door striker could cause the door ajar alarm to function. The technician should closely check all of the doors for the correct adjustment.

Answer B is incorrect. Replacing the door switches without proving they are defective would not be advised.

Answer C is incorrect. Only Technician A is correct.

Answer D is incorrect. Technician A is correct.

TASK E.5

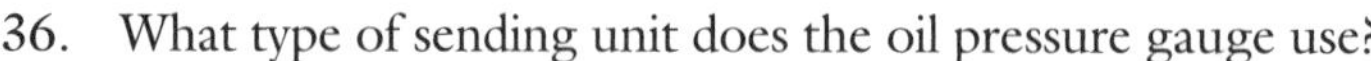

36. What type of sending unit does the oil pressure gauge use?

A. Rheostat
B. Potentiometer
C. Photo resistor
D. Piezo resistor

Answer A is incorrect. A rheostat is sometimes used as the sending unit for the fuel gauge.

Answer B is incorrect. A potentiometer is sometimes used as the sending unit for the fuel gauge.

Answer C is incorrect. A photo resistor is used as an input to automatic light systems as well as some automatic temperature control systems.

Answer D is correct. A piezo resistor is often used as the sending unit for an oil pressure gauge.

TASK E.5

37. Which of the following devices is LEAST LIKELY to be used as an input to an electronic gauge assembly?

A. Thermistor
B. Piezo resistor
C. Body control module
D. Light emitting diode (LED)

Answer A is incorrect. A thermistor is a typical input to a temperature gauge used in an electronic gauge assembly.

Answer B is incorrect. A piezo resistor is a typical input an oil pressure gauge used in an electronic gauge assembly.

Answer C is incorrect. The body control module (BCM) is sometimes used as an input to an electronic gauge assembly.

Answer D is correct. An LED would be considered an output device because it is a type of lamp.

38. The temperature light does not illuminate during the bulb test or when the vehicle overheats. Technician A says that a faulty instrument cluster unit could be the cause. Technician B says that a blown bulb could be the cause. Who is correct?

TASKS E.2, E.4

A. A only
B. B only
C. Both A and B
D. Neither A nor B

Answer A is incorrect. Technician B is also correct.

Answer B is incorrect. Technician A is also correct.

Answer C is correct. Both Technicians are correct. A faulty instrument cluster unit as well as a blown bulb could cause an inoperative temperature light on a late model vehicle.

Answer D is incorrect. Both Technicians are correct.

39. Technician A says that the body control module (BCM) communicates with electronic driver information centers. Technician B says that the power train control module (PCM) communicates information to the electronic driver information center. Who is correct?

TASK E.5

A. A only
B. B only
C. Both A and B
D. Neither A nor B

Answer A is incorrect. Technician B is also correct.

Answer B is incorrect. Technician A is also correct.

Answer C is correct. Both Technicians are correct. Both the body control module (BCM) and the power train control module (PCM) communicate with the electronic driver information center. These modules use the data bus network to send and receive data messages from other controllers. The electronic driver information center communicates on the data bus network.

Answer D is incorrect. Both Technicians are correct.

40. An electronic chime control module needs to be replaced on a late model vehicle. Technician A says that these modules are located in the engine compartment. Technician B says that the body control module (BCM) has to be reprogrammed after replacing the chime control module. Who is correct?

TASK E.5

A. A only
B. B only
C. Both A and B
D. Neither A nor B

Answer A is incorrect. Electronic chime control modules would not be located in the engine compartment.

Answer B is incorrect. Replacing the chime control module would not typically cause the body control module (BCM) to need to be reprogrammed.

Answer C is incorrect. Neither Technician is correct.

Answer D is correct. Neither Technician is correct. Electronic chime control modules are usually located in the passenger compartment. The body control module (BCM) does not typically need to be reprogrammed during the chime module replacement.

TASK F.9

41. A vehicle has a problem with the delay windshield wiper operation. The wipers work on low speed and high speed but do not work at all when the delay function is chosen. What is the most likely cause of this problem?

    A. Washer motor
    B. Potentiometer in the delay switch
    C. Wiper motor low speed brush
    D. Wiper motor high speed brush

    Answer A is incorrect. A washer motor problem would not cause a delay system problem.

    Answer B is correct. A faulty delay wiper switch could cause the delay wiper system to not function correctly.

    Answer C is incorrect. A low speed brush problem would show up on the low wiper speed.

    Answer D is incorrect. A high speed brush problem would show up on the high wiper speed.

TASK F.10

42. A vehicle is having the wiper motor replaced. All of the following steps will need to be performed EXCEPT:

    A. Remove the wiper arms.
    B. Remove the cowl panel.
    C. Remove the wiper motor fasteners.
    D. Remove the vehicle battery from the engine compartment.

    Answer A is incorrect. The wiper arms have to be removed on most vehicles during wiper motor replacement.

    Answer B is incorrect. The cowl panel has to be removed on most vehicles during wiper motor replacement.

    Answer C is incorrect. The wiper motor fasteners always have to be removed during wiper motor replacement.

    Answer D is correct. The vehicle battery does not typically have to be removed during wiper motor replacement.

TASK F.9

43. The windshield washer pump motor runs continuously while the ignition switch is on. Technician A says that the multi-function switch could be shorted. Technician B says that the wiper control module could be defective. Who is correct?

    A. A only
    B. B only
    C. Both A and B
    D. Neither A nor B

    Answer A is incorrect. Technician B is also correct.

    Answer B is incorrect. Technician A is also correct.

    Answer C is correct. Both Technicians are correct. A washer motor that runs continuously could be caused by a shorted multi-function switch, because the wiper switches are incorporated into that switch on several late model vehicles. A defective wiper control module could cause the washer motor to run continuously on some late model vehicles.

    Answer D is incorrect. Both Technicians are correct.

44. A vehicle with heated seats has a problem on the driver's side. The heated seat only gets mildly warm when the switch is turned on. Technician A says that the driver's side heated seat relay could have burnt contacts. Technician B says the driver's side heated seat relay could have an open coil. Who is correct?

TASK F.3

A. A only
B. B only
C. Both A and B
D. Neither A nor B

Answer A is correct. Only Technician A is correct. A relay with burnt contacts could be a point of unwanted resistance in the heated seat circuit. This unwanted resistance could cause the heated seat to not get hot enough.

Answer B is incorrect. An open coil would cause the driver's heat seat to be totally inoperative.

Answer C is incorrect. Only Technician A is correct.

Answer D is incorrect. Technician A is correct.

45. One section of a rear defogger does not clear the glass when in operation. The rest of the glass gets cleared normally when the rear defogger is used. Technician A says that the rear defogger relay could be defective. Technician B says that one of the heat strips could have a break in it. Who is correct?

TASK F.3

A. A only
B. B only
C. Both A and B
D. Neither A nor B

Answer A is incorrect. A defective relay would cause the rear defogger to be weak on all sections of the glass or to be totally inoperative.

Answer B is correct. Only Technician B is correct. A break in one of the heat strips on a rear defogger could cause a small section to not work correctly.

Answer C is incorrect. Only Technician B is correct.

Answer D is incorrect. Technician B is correct.

46. The power door lock only works intermittently on the passenger front door. All of the other doors work normally. Technician A says that the wires could be damaged at the point near where the door connects to the vehicle body. Technician B says that the passenger door lock actuator could have a poor electrical connection. Who is correct?

TASKS F.1, F.2

A. A only
B. B only
C. Both A and B
D. Neither A nor B

Answer A is incorrect. Technician B is also correct.

Answer B is incorrect. Technician A is also correct.

Answer C is correct. Both Technicians are correct. A damaged wire at the point where it passes from the body to the door could cause a power lock problem. A bad connection at the passenger door lock actuator could cause an intermittent problem too.

Answer D is incorrect. Both Technicians are correct.

TASK F.6

47. The keyless entry control module is being replaced on a late-model vehicle. Technician A says that all of the transmitters will work with the new controller without having to be reprogrammed. Technician B says that the power door locks will have to be cycled multiple times to reprogram the keyless entry control module. Who is correct?

A. A only

B. B only

C. Both A and B

D. Neither A nor B

Answer A is incorrect. The old transmitters will need to be reprogrammed into the new keyless entry module in order for the system to work correctly.

Answer B is incorrect. Cycling the power door locks does not reprogram the keyless entry module.

Answer C is incorrect. Neither Technician is correct.

Answer D is correct. Neither Technician is correct. The old keyless entry transmitters will need to be programmed into the new module before they will work. The keyless entry system will have to be put into programming mode to get the transmitters to work correctly. This process will vary among the vehicle manufacturers.

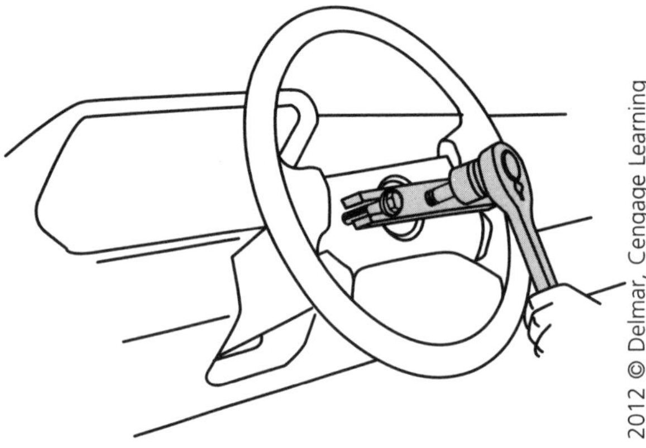

TASK F.10

48. Which cruise control system component is most likely being serviced in the figure above?

A. Clock spring/ribbon wire

B. Servo

C. Control module

D. Cancel switch

Answer A is correct. The steering wheel is being removed in order to replace the clock spring. The cruise control circuit is routed through the clock spring when the cruise control switches are located on the steering wheel.

Answer B is incorrect. The cruise servo is located in the engine compartment.

Answer C is incorrect. The cruise control module is located under the dash area.

Answer D is incorrect. The cancel switch is located on the horn pad or near the steering column area.

49. Technician A says that audio speakers can be checked by testing the resistance with an ohmmeter. Technician B says that an antenna lead wire can be checked by testing the resistance with an ohmmeter. Who is correct?

TASK F.8

A. A only

B. B only

C. Both A and B

D. Neither A nor B

Answer A is incorrect. Technician B is also correct.

Answer B is incorrect. Technician A is also correct.

Answer C is correct. Both Technicians are correct. An ohmmeter can be used to test the audio speakers as well as to test the antenna lead wire.

Answer D is incorrect. Both Technicians are correct.

50. A digital clock is being diagnosed for losing the correct time each time the ignition is shut off. Technician A says that the main ground connection for the clock could be disconnected. Technician B says that the "keep alive" fuse could be blown. Who is correct?

TASKS F.1, F.7

A. A only

B. B only

C. Both A and B

D. Neither A nor B

Answer A is incorrect. The digital clock would be totally inoperative if the main ground connection was disconnected.

Answer B is correct. Only Technician B is correct. A blown "keep alive" fuse could cause the clock to lose its time each time the ignition switch is turned off.

Answer C is incorrect. Only Technician B is correct.

Answer D is incorrect. Technician B is correct.

PREPARATION EXAM 3—ANSWER KEY

1. C	**21.** D	**41.** B
2. B	**22.** C	**42.** C
3. B	**23.** C	**43.** B
4. B	**24.** C	**44.** A
5. B	**25.** C	**45.** B
6. B	**26.** B	**46.** D
7. C	**27.** B	**47.** B
8. C	**28.** C	**48.** C
9. A	**29.** D	**49.** D
10. A	**30.** B	**50.** A
11. C	**31.** D	
12. D	**32.** D	
13. C	**33.** C	
14. C	**34.** C	
15. A	**35.** D	
16. D	**36.** A	
17. C	**37.** B	
18. D	**38.** A	
19. A	**39.** C	
20. B	**40.** B	

PREPARATION EXAM 3—EXPLANATIONS

TASKS
A.1, D.3, F.8

1. Which of the following practices would be the LEAST LIKELY use of a 12 volt test light?

   A. Checking for power at a rear defogger grid
   B. Checking for power at a headlight bulb
   C. Checking for reference voltage at temperature sending unit
   D. Checking the input and output voltage of a fuse at the fuse panel

Answer A is incorrect. A test light could be used to check for power at a rear defogger grid.

Answer B is incorrect. A test light could be used to test for power at a headlight bulb.

Answer C is correct. A test light would not be a safe tool to check for reference voltage at a temperature sending unit. A high impedance multimeter should be used to test electronic circuits.

Answer D is incorrect. A test light could be used to test input and output voltages of the fuses.

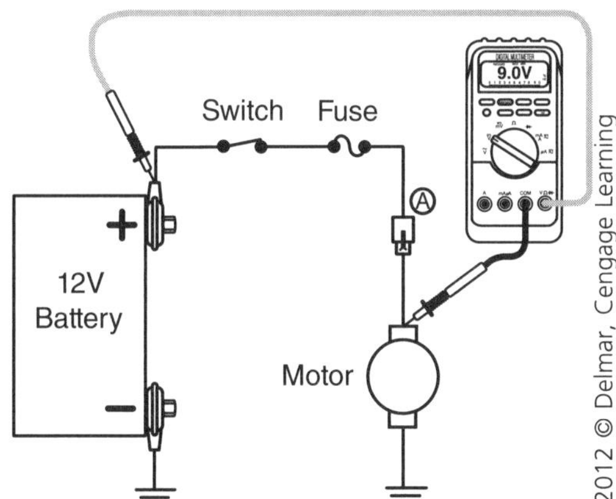

2. Referring to the 12 volt circuit above, the motor runs slowly when the switch is turned on. Technician A says that a faulty motor ground could cause this condition. Technician B says that faulty terminal at connector A could be the cause. Who is correct?

TASK A.1

A. A only
B. B only
C. Both A and B
D. Neither A nor B

Answer A is incorrect. The reading shows that 9 volts is being dropped on the power side of the circuit before the load which indicates that the problem is on that side of the circuit.

Answer B is correct. Only Technician B is correct. A faulty terminal in connector A could cause the excessive voltage shown in the figure. The voltage drop on the power side of the circuit should be very low. The motor should drop nearly 12 volts if everything is operating correctly.

Answer C is incorrect. Only Technician B is correct.

Answer D is incorrect. Technician B is correct.

3. Technician A says that a poor ground connection will cause increased current flow in an electrical circuit which would likely blow a fuse. Technician B says that water corrosion in the wiring will cause reduced current flow in an electrical circuit. Who is correct?

TASK A.2

A. A only
B. B only
C. Both A and B
D. Neither A nor B

Answer A is incorrect. A poor ground will reduce current flow in an electrical circuit.

Answer B is correct. Only Technician B is correct. Corrosion in electrical wiring would cause increased electrical resistance which would reduce the current flow.

Answer C is incorrect. Only Technician B is correct.

Answer D is incorrect. Technician B is correct.

TASK A.2

4. Which of the following methods of testing amperage in a circuit would be LEAST LIKELY to interrupt the circuit?

   A. Remove the switch and connect the ammeter leads to the exposed terminals
   B. Connecting an amp clamp around the wire
   C. Remove the relay and connect the ammeter leads across the control side of the circuit
   D. Connecting the voltmeter across the fuse

Answer A is incorrect. Removing the switch would require the circuit to be interrupted. However, if an amp clamp is not available, then the switch is sometimes a good place to measure current.

Answer B is correct. Using an inductive amp clamp does not require that the circuit be interrupted at all. These tools are very functional on late model cars and trucks.

Answer C is incorrect. Removing the relay would cause some interruption of the circuit. However, the relay is a good location to test current if an amp clamp is not available.

Answer D is incorrect. Connecting the voltmeter across the fuse would not measure current in the circuit.

TASK A.3

5. A resistance test was performed on an open headlight switch and the result is 3.857 megohms. Technician A says that the switch is faulty because the reading is beyond the specifications. Technician B says that the switch has over 3 million ohms of resistance. Who is correct?

   A. A only
   B. B only
   C. Both A and B
   D. Neither A nor B

Answer A is incorrect. An open switch is supposed to have very high resistance.

Answer B is correct. Only Technician B is correct. The reading is greater than 3 million ohms. An open switch should have very high resistance.

Answer C is incorrect. Only Technician B is correct.

Answer D is incorrect. Technician B is correct.

TASK A.4

6. Technician A says that an oscilloscope can be used as a scan tool to read live data and codes. Technician B says that an oscilloscope can be used to view live sensor readings directly at each sensor. Who is correct?

   A. A only
   B. B only
   C. Both A and B
   D. Neither A nor B

Answer A is incorrect. Oscilloscopes display voltage over time on a digital screen, but they do not read data and codes. However, some scan tools have oscilloscopes built into them.

Answer B is correct. Only Technician B is correct. Oscilloscopes can be used to view live sensor readings directly at each sensor.

Answer C is incorrect. Only Technician B is correct.

Answer D is incorrect. Technician B is correct.

7. Technician A says that a scan tool can be used to retrieve a trouble code from the engine control module. Technician B says that a scan tool can be used to view live sensor data from the engine control module. Who is correct?

TASK A.5

A. A only
B. B only
C. Both A and B
D. Neither A nor B

Answer A is incorrect. Technician B is also correct.

Answer B is incorrect. Technician A is also correct.

Answer C is correct. Both Technicians are correct. Scan tools can be used to retrieve trouble codes as well as view live sensor data from vehicle control modules.

Answer D is incorrect. Both Technicians are correct.

8. Technician A says that a fused jumper wire can be used to provide power to the horn during diagnosis. Technician B says that a fused jumper wire can be used to bypass a rear defogger relay during the diagnosis of that circuit. Who is correct?

TASKS F.4, F.10

A. A only
B. B only
C. Both A and B
D. Neither A nor B

Answer A is incorrect. Technician B is also correct.

Answer B is incorrect. Technician A is also correct.

Answer C is correct. Both Technicians are correct. Fused jumper wires can be used to bypass relays as well as supply direct power to electrical items during diagnosis.

Answer D is incorrect. Both Technicians are correct.

9. Technician A says that a wire that is shorted to ground before the load will cause a fuse to blow due to high current flow. Technician B says that a corroded wire connection will cause the electrical load to receive increased current flow. Who is correct?

TASK A.6

A. A only
B. B only
C. Both A and B
D. Neither A nor B

Answer A is correct. Only Technician A is correct. A short to ground before the load will blow the fuse due to high current flow.

Answer B is incorrect. Corrosion in electrical wires will reduce the current flow that is supplied to the electrical load.

Answer C is incorrect. Only Technician A is correct.

Answer D is incorrect. Technician A is correct.

TASK A.7

10. All of the following procedures are acceptable methods of locating excessive parasitic draw EXCEPT:

A. Disconnect the battery ground cable from the engine block.

B. Remove the fuses one at a time while watching the ammeter.

C. Inspect the whole vehicle for any lamp that could be staying on.

D. Disconnect the charge wire at the generator while watching the ammeter.

Answer A is correct. Disconnecting the negative battery cable at the engine block will not assist in finding an excessive parasitic draw. All current flow will stop when the battery cable is removed.

Answer B is incorrect. Removing the fuses one at a time is an organized way to find an excessive parasitic draw.

Answer C is incorrect. Inspecting the vehicle for lamps that are staying on is a good step in diagnosing an excessive parasitic draw.

Answer D is incorrect. Disconnecting the charge wire at the generator is a step in diagnosing an excessive parasitic draw problem.

TASK A.8

11. Which of the following conditions would LEAST LIKELY cause a circuit breaker to open?

A. A stalled motor

B. A power-side short to ground

C. A ground-side short to ground

D. A motor with a shorted winding

Answer A is incorrect. A stalled motor could cause a circuit breaker to open due to the rapid increase in current flow.

Answer B is incorrect. A power-side short to ground could cause a circuit breaker to open due to the rapid increase in current flow.

Answer C is correct. A ground-side short to ground will not cause a fuse to open. If the circuit is ground-side switched, then the circuit will not turn off.

Answer D is incorrect. A motor with a shorted winding could cause a circuit breaker to open due to an increase in current flow.

TASK A.9

12. All of the following items are typically found in a wiring diagram EXCEPT:

A. The color of the wire

B. The circuit number of a wire

C. The connector numbers for the circuit in question

D. The locations of each splice used in the circuit

Answer A is incorrect. Wiring diagrams will usually show the colors of the wires involved in the circuit.

Answer B is incorrect. Wiring diagrams will usually show the circuit numbers of the wires involved in the circuit.

Answer C is incorrect. Wiring diagrams usually show the connector numbers as well as pins identifications of the wires involved in the circuit.

Answer D is correct. Wiring diagrams do not typically describe the location of the splices that are used in the circuit.

13. Technician A says that a high impedance digital meter is needed to perform voltage tests on the data bus network. Technician B says that an oscilloscope can be used to view the communication activity on the data bus network. Who is correct?

TASK A.11

A. A only
B. B only
C. Both A and B
D. Neither A nor B

Answer A is incorrect. Technician B is also correct.

Answer B is incorrect. Technician A is also correct.

Answer C is correct. Both Technicians are correct. Testing on the data bus network can be done with high impedance digital meters. In addition, an oscilloscope can be used to view the data bus wires for electrical activity.

Answer D is incorrect. Both Technicians are correct.

14. Technician A says that the battery voltage should be 12.4 volts before performing a load test. Technician B says that a battery state of charge can be determined by measuring the open circuit voltage across the terminals. Who is correct?

TASKS B.1, B.2

A. A only
B. B only
C. Both A and B
D. Neither A nor B

Answer A is incorrect. Technician B is also correct.

Answer B is incorrect. Technician A is also correct.

Answer C is correct. Both Technicians are correct. A battery needs to be at least 75 percent charged before a valid load test can be performed. A battery with 12.4 volts is 75 percent charged. A fully charged battery will measure 12.6 volts during the open circuit test.

Answer D is incorrect. Both Technicians are correct.

15. A battery is being replaced in a vehicle. Technician A says that the replacement battery should have at least the minimum CCA rating for the vehicle. Technician B says that the replacement battery should not exceed the reserve capacity of the original battery. Who is correct?

TASK B.5

A. A only
B. B only
C. Both A and B
D. Neither A nor B

Answer A is correct. Only Technician A is correct. Replacement batteries should have at least the minimum CCA rating set for the vehicle.

Answer B is incorrect. A replacement battery can have a higher reserve capacity without causing any problems for the vehicle.

Answer C is incorrect. Only Technician A is correct.

Answer D is incorrect. Technician A is correct.

TASK B.4

16. Technician A says that batteries can be recharged more quickly by using a low setting on the battery charger. Technician B says that batteries can be more thoroughly charged by using a high setting on the battery charger. Who is correct?

A. A only
B. B only
C. Both A and B
D. Neither A nor B

Answer A is incorrect. Slow charging vehicle batteries takes more time than fast charging. However, slow charging is usually more thorough than fast charging.

Answer B is incorrect. Slow charging will more thoroughly recharge a battery than fast charging them.

Answer C is incorrect. Neither Technician is correct.

Answer D is correct. Neither Technician is correct. Fast charging vehicle batteries is quicker but slow charging vehicle batteries is more thorough.

TASK B.6

17. A vehicle needs to be jumpstarted with an auxiliary power supply. Technician A says that safe practices should always be followed when working around batteries. Technician B says that the negative connection should be made at a secure engine ground. Who is correct?

A. A only
B. B only
C. Both A and B
D. Neither A nor B

Answer A is incorrect. Technician B is also correct.

Answer B is incorrect. Technician A is also correct.

Answer C is correct. Both Technicians are correct. Extreme care should be practiced when working around batteries. It is a good practice to connect the negative cable to a secure engine ground rather than at the battery. Doing this will reduce the likelihood of causing a battery explosion.

Answer D is incorrect. Both Technicians are correct.

TASK B.7

18. Technician A says that burnt starter solenoid contacts can cause elevated starter current draw. Technician B says that a partially cut battery cable can cause elevated starter current draw. Who is correct?

A. A only
B. B only
C. Both A and B
D. Neither A nor B

Answer A is incorrect. Burnt solenoid contacts will decrease the starter current draw due to the increased electrical resistance.

Answer B is incorrect. A partially cut battery cable will cause a decrease in starter current draw due to the decreased area for electrons to flow.

Answer C is incorrect. Neither Technician is correct.

Answer D is correct. Neither Technician is correct. Burn solenoid contacts and a damaged battery cable will both reduce starter current draw. These two conditions will both cause the electrical resistance to rise, which will reduce the current flow.

19. Which of the following components is LEAST LIKELY to be part of the starter control circuit.

TASK B.9

A. Positive battery cable
B. Ignition switch
C. Park/neutral switch
D. Starter relay

Answer A is correct. The positive battery cable is directly connected to the starter solenoid and provides power to the load (high amperage) side of the starting circuit.

Answer B is incorrect. The ignition switch is the device that allows the driver to engage the start sequence. It is part of the control circuit.

Answer C is incorrect. The park/neutral switch prevents the starter from engaging while in gear. It is part of the control circuit.

Answer D is incorrect. The starter relay is in the starter control circuit. This device receives power from the ignition switch and then sends power to the control side of the solenoid.

20. Technician A says that the starter solenoid contacts can be tested by performing a voltage drop test across the "bat" and "motor" terminals when the starter is disengaged. Technician B says that the pull-in winding can be tested by measuring the resistance from the "S" terminal to the "motor" terminal. Who is correct?

TASK B.9

A. A only
B. B only
C. Both A and B
D. Neither A nor B

Answer A is incorrect. The starter contacts must be tested with the voltage drop test while the starter is engaged.

Answer B is correct. Only Technician B is correct. The pull-in winding can be tested with an ohmmeter connected from the "S" terminal to the "motor" terminal. This resistance is typically about 1 ohm.

Answer C is incorrect. Only Technician B is correct.

Answer D is incorrect. Technician B is correct.

21. Technician A says that the airbag fuse should be removed prior to disconnecting the electrical connections at the starter. Technician B says that the starter can be supported by the electrical wires without any expected damage to the wires. Who is correct?

TASK B.9

A. A only
B. B only
C. Both A and B
D. Neither A nor B

Answer A is incorrect. It is not necessary to remove the airbag fuse when unhooking the starter.

Answer B is incorrect. It is not acceptable to let the starter hang by the electrical wires.

Answer C is incorrect. Neither Technician is correct.

Answer D is correct. Neither Technician is correct. The negative battery cable should be disconnected prior to disconnecting the electrical wires at the starter. The starter should never be supported by the electrical wires alone.

TASK B.10

22. Technician A says that a mistimed engine can cause slow cranking speed. Technician B says that low engine compression can cause rapid cranking speed. Who is correct?

A. A only
B. B only
C. Both A and B
D. Neither A nor B

Answer A is incorrect. Technician B is also correct.

Answer B is incorrect. Technician A is also correct.

Answer C is correct. Both Technicians are correct. An engine that is mistimed will sometimes crank slower than normal due to the ignition system firing at the wrong time. In addition, an engine that has low compression will usually have a rapid cranking speed.

Answer D is incorrect. Both Technicians are correct.

TASK C.1

23. Technician A says that a poor connection at the charging output wire can cause a charging problem. Technician B says that many charging circuits contain a circuit protection device. Who is correct?

A. A only
B. B only
C. Both A and B
D. Neither A nor B

Answer A is incorrect. Technician B is also correct.

Answer B is incorrect. Technician A is also correct.

Answer C is correct. Both Technicians are correct. A poor connection at the charging output wire will cause an unwanted voltage drop, which will reduce the charging capacity to the battery. Most charging circuits have either a fusible link or a large fuse that protects that circuit in case of a grounding problem.

Answer D is incorrect. Both Technicians are correct.

TASK C.2

24. The accessory drive belt system should be inspected during regular intervals. Technician A says that a serpentine drive belt tensioner should maintain spring pressure on the belt at all times. Technician B says that the drive belt should be replaced if major pieces are missing from the drive surface. Who is correct?

A. A only
B. B only
C. Both A and B
D. Neither A nor B

Answer A is incorrect. Technician B is also correct.

Answer B is incorrect. Technician A is also correct.

Answer C is correct. Both Technicians are correct. The belt tensioner contains a strong spring and should maintain constant pressure on the belt at all times. A serpentine drive belt should be replaced if big pieces are missing from the drive surface.

Answer D is incorrect. Both Technicians are correct.

TASKS C.2, C.4

25. A vehicle is being diagnosed for a charging problem. The generator produced 50 amps during the output test and it is rated at 105 amps. Technician A says that the generator could have a worn brushes. Technician B says that the generator drive belt could be slipping. Who is correct?

A. A only

B. B only

C. Both A and B

D. Neither A nor B

Answer A is incorrect. Technician B is also correct.

Answer B is incorrect. Technician A is also correct.

Answer C is correct. Both Technicians are correct. Worn brushes can cause low charging output due to not creating enough field current in the rotor. A slipping drive belt can cause low charging output and would likely create a loud noise to alert the technician of the problem.

Answer D is incorrect. Both Technicians are correct.

TASK C.7

26. All of the following methods of wire repair in the charging system are currently used EXCEPT:

A. Crimp and seal connectors

B. Butt connectors and tape

C. New wiring harness

D. Solder and heat shrink

Answer A is incorrect. Crimp and seal connectors are acceptable to use in charging system repairs. These connectors are sealed against water intrusion.

Answer B is correct. Using butt connectors and tape is not a professional wire repair practice because it will not prevent water from entering the connection.

Answer C is incorrect. A new wiring harness is sometimes necessary to satisfy some manufacturers' requirements.

Answer D is incorrect. Solder and heat shrink methods of wire repair are acceptable for charging system repairs. This repair technique does not allow water intrusion.

TASK C.8

27. A generator is being replaced. Technician A uses an air tool to install the fastening nut onto the charging output wire. Technician B carefully routes the drive belt around all of the pulleys before releasing the belt tensioner. Who is correct?

A. A only

B. B only

C. Both A and B

D. Neither A nor B

Answer A is incorrect. An air tool should never be used to install the fastening nut onto the charging output wire. This would likely cause internal damage to the generator.

Answer B is correct. Only Technician B is correct. The drive belt should be routed around all of the drive pulleys before releasing the tensioner.

Answer C is incorrect. Only Technician B is correct.

Answer D is incorrect. Technician B is correct.

TASKS
D.1, D.3

28. The dimmer switch has failed and needs to be replaced on a late-model vehicle. Technician A says that some dimmer switches are built into the multi-function switch. Technician B says that replacing some dimmer switches requires the removal of the steering wheel. Who is correct?

    A. A only
    B. B only
    C. Both A and B
    D. Neither A nor B

    Answer A is incorrect. Technician B is also correct.

    Answer B is incorrect. Technician A is also correct.

    Answer C is correct. Both Technicians are correct. Many late model vehicles have combined the dimmer switch into the multi-function switch. Replacing some multi-function switches requires the removal of the steering wheel.

    Answer D is incorrect. Both Technicians are correct.

TASKS
E.1, E.4

29. Which of the following devices regulates the brightness of the dash lights?

    A. Positive temperature coefficient
    B. Door jam switch
    C. Dimmer switch
    D. Rheostat

    Answer A is incorrect. A positive temperature coefficient is a thermistor that increases its resistance as it heats up. These devices are sometimes used as circuit protection devices.

    Answer B is incorrect. The door jam switch causes the dome lights to illuminate when a door is opened.

    Answer C is incorrect. The dimmer switch changes the headlights levels to either dim or bright.

    Answer D is correct. A rheostat is typically used to control the brightness of the dashlights.

TASK D.6

30. The courtesy lights stay on continuously and have caused the battery to be discharged while the vehicle is parked for long periods of time. Technician A says that the courtesy light switch could have a bad connection. Technician B says that a door ajar switch could be shorted. Who is correct?

    A. A only
    B. B only
    C. Both A and B
    D. Neither A nor B

    Answer A is incorrect. A bad connection at the courtesy light switch would likely cause the lights to be dim or possibly totally inoperative.

    Answer B is correct. Only Technician B is correct. A shorted door jam switch could cause the courtesy lights to stay on continuously.

    Answer C is incorrect. Only Technician B is correct.

    Answer D is incorrect. Technician B is correct.

31. All of the following conditions could cause the brake lights to be inoperative EXCEPT:

TASKS D.4, E.4

A. Brake switch is stuck "open."

B. Brake light fuse is blown.

C. Brake light bulbs are blown.

D. Battery cable is loose.

Answer A is incorrect. A brake switch that is stuck "open" would cause the brake lights to not work correctly.

Answer B is incorrect. A brake fuse that is blown would cause the brake lights to be inoperative.

Answer C is incorrect. Blown brake light bulbs could cause inoperative brake lights.

Answer D is correct. A loose battery cable would not cause inoperative brake lights. A loose cable could cause the starter to crank slowly or possibly a charging problem.

32. A vehicle is being diagnosed for inoperative turn signals. All of the hazard lights work correctly. Technician A says that a blown turn signal bulb could be the cause. Technician B says that an open ground connection at the right rear lamp socket could be the cause. Who is correct?

TASK D.4

A. A only

B. B only

C. Both A and B

D. Neither A nor B

Answer A is incorrect. The turn signals and the hazard lights use the same bulbs so a blown bulb would not be the cause of this problem.

Answer B is incorrect. The turn signals and the hazard lights use the same bulbs so an open ground at the right rear lamp socket would not be the cause of this problem.

Answer C is incorrect. Neither Technician is correct.

Answer D is correct. Neither Technician is correct. This problem has to be something that is common only to the turn signal system. Both of the examples would have caused the hazard lights to not burn correctly.

33. A vehicle needs to have a trailer wiring harness installed into the existing lighting harness. Technician A says that many trailer wiring companies provide harnesses that tee directly into the existing harness. Technician B says that the turn signal flasher will need to be replaced with a heavy duty flasher. Who is correct?

TASK D.8

A. A only

B. B only

C. Both A and B

D. Neither A nor B

Answer A is incorrect. Technician B is also correct.

Answer B is incorrect. Technician A is also correct.

Answer C is correct. Both Technicians are correct. Many wiring companies sell harnesses that just plug into the existing harness. This prevents the need to splice into the harness of the existing wires. The turn signal flasher will need to be replaced with a heavy duty flasher so the flash rate will not be affected when the vehicle is connected to a trailer.

Answer D is incorrect. Both Technicians are correct.

TASK E.4

34. A vehicle is being diagnosed with a problem of the door ajar alarm sounding intermittently. The vehicle is driven for several miles without duplicating the problem. Technician A says that the striker adjustment of all doors should be checked. Technician B says that the door jam switch adjustment should be checked. Who is correct?

A. A only
B. B only
C. Both A and B
D. Neither A nor B

Answer A is incorrect. Technician B is also correct.

Answer B is incorrect. Technician A is also correct.

Answer C is correct. Both Technicians are correct. It would be advisable to check both the door striker adjustments as well as the door switch adjustments to try to locate this problem. Another possible way to try to duplicate this problem would be to lift the vehicle on a vehicle hoist and see if the problem showed up as the vehicle body is flexed.

Answer D is incorrect. Both Technicians are correct.

TASK E.3

35. A vehicle is being diagnosed with a fuel gauge that does not work. Technician A says that the fuel tank should be removed to test the fuel sending unit. Technician B says that the instrument cluster should be removed to inspect the connections. Who is correct?

A. A only
B. B only
C. Both A and B
D. Neither A nor B

Answer A is incorrect. The fuel tank does not have to be removed to test the fuel sending unit and related circuit.

Answer B is incorrect. The instrument cluster does not need to be removed as the first step in the diagnosis. Some clusters will be very labor intensive to remove them.

Answer C is incorrect. Neither Technician is correct.

Answer D is correct. Neither Technician is correct. The technician should first verify that the other gauges are working. If the other gauges work, then go to the fuel harness connector near the fuel tank and disconnect it. The fuel gauge should move to full or empty while the connector is unhooked. Next, connect a fused jumper wire between the circuit for the gauge and chassis ground. This step should cause the gauge to read opposite of what it did when it was unhooked. If the gauge moves to both full and empty, then the likely cause of the problem is a faulty sending unit. If the gauge does not respond to the above tests, then further testing will need to be done.

TASKS A.2, E.5

36. What type of sending unit does the fuel gauge use?

A. Rheostat
B. Transducer
C. Photo resistor
D. Thermistor

Answer A is correct. Fuel gauges typically use a rheostat or a potentiometer that connects to a float that rides on the fuel as it rises and falls.

Answer B is incorrect. A transducer is used on systems with varying pressure such as the air conditioning system.

Answer C is incorrect. A photo resistor is used as an input on automatic light systems as well as some automatic temperature control systems.

Answer D is incorrect. A thermistor is used as a sending unit for the temperature gauges as well as an input to measure temperature for several other electronic systems such as the engine, transmission and the HVAC system.

37. A vehicle is being diagnosed for a problem of all of the gauges being inoperative. The vehicle is equipped with an electronic cluster that has self-diagnostic capabilities. Technician A says that every sending unit should be checked individually. Technician B says that this problem could be diagnosed by using the self-diagnostic function of the instrument cluster. Who is correct?

TASK E.3

A. A only
B. B only
C. Both A and B
D. Neither A nor B

Answer A is incorrect. It is not likely that every sending unit could be defective at the same time.

Answer B is correct. Only Technician B is correct. The self-diagnostic feature of the cluster would provide valuable assistance in troubleshooting a problem like this. A scan tool could also be used to retrieve data, check for trouble codes as well as to perform functional tests on the cluster.

Answer C is incorrect. Only Technician B is correct.

Answer D is incorrect. Technician B is correct.

38. The temperature light stays on continuously while driving the vehicle. Technician A says that a shorted wire leading to the temperature sending unit could be the cause. Technician B says that an "open" temperature sending unit could be the cause. Who is correct?

TASKS E.4, E.5

A. A only
B. B only
C. Both A and B
D. Neither A nor B

Answer A is correct. Only Technician A is correct. A shorted wire could cause the temperature light to stay on continuously. The temperature sending unit causes the light to illuminate by providing a ground for the bulb.

Answer B is incorrect. An "open" temperature sending unit would cause the light to not come on at all.

Answer C is incorrect. Only Technician A is correct.

Answer D is incorrect. Technician A is correct.

39. Technician A says that the electronic driver information center receives data from the vehicle data bus network. Technician B says that the electronic driver information center can display "data bus" messages to the driver or technician. Who is correct?

TASK E.5

A. A only
B. B only
C. Both A and B
D. Neither A nor B

Answer A is incorrect. Technician B is also correct.

Answer B is incorrect. Technician A is also correct.

Answer C is correct. Both Technicians are correct. Electronic driver information centers typically receive information from the vehicle data bus network. Some electronic driver information centers have the ability to give "data bus" messages.

Answer D is incorrect. Both Technicians are correct.

TASK F.9

40. A vehicle has a problem with the delay windshield wiper operation. The wipers work on low speed and high speed but do not work at all when the delay function is chosen. Technician A says that the main ground connection for the wiper system could be poorly connected. Technician B says that the wiper control module could be faulty. Who is correct?

    A. A only
    B. B only
    C. Both A and B
    D. Neither A nor B

Answer A is incorrect. A faulty main ground for the wiper system would prevent operation on all speeds.

Answer B is correct. Only Technician B is correct. A faulty wiper control module could cause the delay wipers to not function correctly. The delay portion of the wiper switch could also cause this problem.

Answer C is incorrect. Only Technician B is correct.

Answer D is incorrect. Technician B is correct.

TASK F.10

41. A vehicle is having the wiper motor replaced. Which of the following steps is LEAST LIKELY to be performed?

    A. Remove the wiper arms.
    B. Remove the accessory drive belt.
    C. Remove the wiper motor fasteners.
    D. Remove the cowl panel.

Answer A is incorrect. The wiper arms would need to be removed to replace the wiper motor on many late model vehicles.

Answer B is correct. The drive belt would have nothing to do with the wiper motor.

Answer C is incorrect. The wiper motor fasteners would need to be removed to replace the wiper motor.

Answer D is incorrect. The cowl panel would need to be removed to replace the wiper motor on many late model vehicles.

TASK F.9

42. The windshield washer pump is totally inoperative. Technician A says that the wiper switch could be defective. Technician B says that the wiper control module could be defective. Who is correct?

    A. A only
    B. B only
    C. Both A and B
    D. Neither A nor B

Answer A is incorrect. Technician B is also correct.

Answer B is incorrect. Technician A is also correct.

Answer C is correct. Both Technicians are correct. An inoperative washer could be caused by a defective wiper switch or a defective wiper control module. A bad washer pump could also cause this problem.

Answer D is incorrect. Both Technicians are correct.

43. A vehicle with heated seats has a problem on the passenger side. The heated seat only gets mildly warm when the switch is turned on. Technician A says that the passenger side heated seat relay could have open contacts when turned on. Technician B says the passenger side heated seat heat strip could have a poor connection. Who is correct?

TASK F.3

A. A only
B. B only
C. Both A and B
D. Neither A nor B

Answer A is incorrect. Open contacts on the passenger side heated seat relay would cause theheated seat to be totally inoperative.

Answer B is correct. Only Technician B is correct. A poor connection at the heat strip could cause the seat to not warm up properly.

Answer C is incorrect. Only Technician B is correct.

Answer D is incorrect. Technician B is correct.

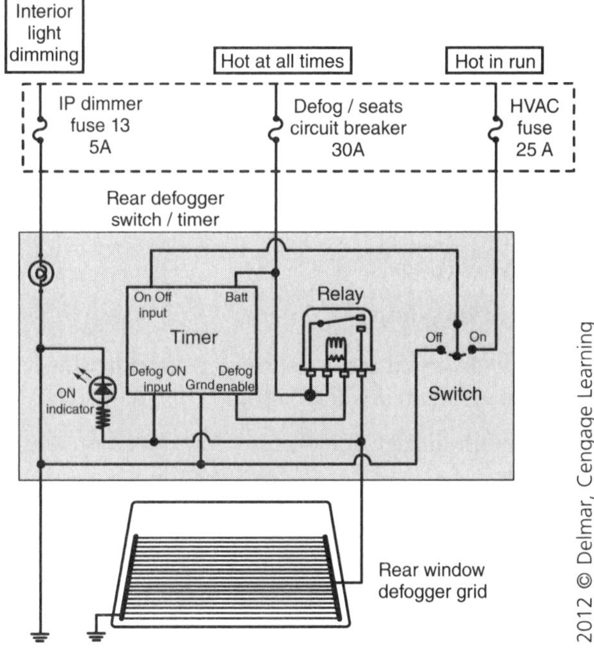

2012 © Delmar, Cengage Learning

44. Referring to the figure above, the rear defogger does not clear the back glass when the system is turned on. The "on indicator" illuminates when the system is turned on. Technician A says that an open ground at the rear defogger grid could be the cause. Technician B says that an open grid line could be the cause. Who is correct?

TASKS F.3, F.4

A. A only
B. B only
C. Both A and B
D. Neither A nor B

Answer A is correct. Only Technician A is correct. An open ground at the rear defogger grid would cause the whole grid to be inoperative. All circuits need to have a quality ground in order to function correctly.

Answer B is incorrect. An open grid line would only cause one section of the grid to be inoperative.

Answer C is incorrect. Only Technician A is correct.

Answer D is incorrect. Technician A is correct.

TASKS
F.4, F.5

45. The power door lock at the driver's door does not work at all. All of the other actuators work normally. Technician A says that the wires could be damaged at the point near where the door connects to the vehicle body. Technician B says that the driver's door lock actuator could be faulty. Who is correct?

A. A only

B. B only

C. Both A and B

D. Neither A nor B

Answer A is incorrect. Since the power locks work at all of the locations except the driver's door, it is very unlikely that broken wires between the body and the door could cause this problem.

Answer B is correct. Only Technician B is correct. A faulty driver's door lock actuator could cause the power locks to be inoperative at one door and still work correctly at the other doors.

Answer C is incorrect. Only Technician B is correct.

Answer D is incorrect. Technician B is correct.

TASKS
F.1, F.2

46. All of the following conditions could cause the sunroof to operate slowly EXCEPT:

A. Burnt relay contacts

B. Tight sunroof motor linkage

C. Binding sunroof tracks

D. Using a 40 amp fuse in place of a 30 amp fuse

Answer A is incorrect. Burnt relays contacts could cause a sunroof to operate slowly because the burnt contacts would likely cause an unwanted voltage drop.

Answer B is incorrect. Tight sunroof motor linkage could cause slow operation due to the added resistance to physical movement.

Answer C is incorrect. Binding sunroof tracks could cause slow operation due the added resistance to physical movement.

Answer D is correct. Having a fuse that is rated too high would not cause the sunroof to operate slowly. A technician should never install any type of circuit protection device that is larger than the specifications dictate.

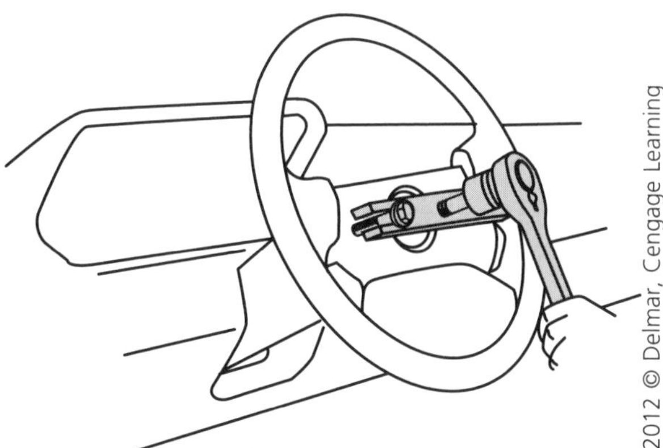

47. Which airbag system component is most likely being serviced in the figure above?

TASKS F.9, F.10

A. Crash sensor

B. Clock spring

C. Control module

D. Arming sensor

Answer A is incorrect. The crash sensors for the airbag system are typically located near the front of the vehicle.

Answer B is correct. The clock spring is located under the steering wheel. This device has a coiled wire that provides a "hard wire" connection to the airbag inflator. The other steering wheel circuits such as the horn, cruise, radio and temperature control are also wired through the clock spring.

Answer C is incorrect. The control module for the airbag system is not located in the steering column. The control module is usually located under the dash area or on the floor area of the vehicle.

Answer D is incorrect. The arming sensor for the airbag system is not located in the steering column. The arming sensor is usually located in the floor area of the vehicle.

48. Technician A says that some power antenna systems utilize a relay to raise the power antenna when the radio is turned on. Technician B says that some power antenna systems utilize a relay to lower the power antenna when the radio is turned off. Who is correct?

TASK F.7

A. A only

B. B only

C. Both A and B

D. Neither A nor B

Answer A is incorrect. Technician B is also correct.

Answer B is incorrect. Technician A is also correct.

Answer C is correct. Both Technicians are correct. The power antenna system typically uses relays to raise and lower the power antenna when the radio is turned on and off. Some manufacturers have a switch that allows the antenna to be raised and lowered.

Answer D is incorrect. Both Technicians are correct.

TASK F.9

49. The cruise control is inoperative on a late-model vehicle. Technician says that a broken speedometer cable is the likely cause. Technician B says that a misadjusted throttle plate could be the cause. Who is correct?

A. A only

B. B only

C. Both A and B

D. Neither A nor B

Answer A is incorrect. Most late-model vehicles do not use speedometer cables. Speed sensors are used to generate a variable signal that that is sent to a computer as a speed input.

Answer B is incorrect. A misadjusted throttle plate would not likely cause an inoperative cruise control system.

Answer C is incorrect. Neither Technician is correct.

Answer D is correct. Neither Technician is correct. Speed sensors are used on most late-model vehicles to provide speed input to the engine, transmission and cruise control systems. A misadjusted throttle plate would not likely cause an inoperative cruise control system.

TASK F.5

50. A vehicle with a theft alarm system is being diagnosed for a problem of the alarm will not activate. Technician A says that a misadjusted door could be the cause. Technician B says that a faulty ignition switch could be the cause. Who is correct?

A. A only

B. B only

C. Both A and B

D. Neither A nor B

Answer A is correct. Only Technician A is correct. A misadjusted door could cause the alarm system to fail to operate correctly. All of the door adjustments should be checked during the diagnosis of this problem.

Answer B is incorrect. A faulty ignition switch would not likely cause a problem with the theft alarm system.

Answer C is incorrect. Only Technician A is correct.

Answer D is incorrect. Technician A is correct.

PREPARATION EXAM 4—ANSWER KEY

1. D
2. B
3. A
4. B
5. C
6. C
7. C
8. C
9. A
10. D
11. B
12. A
13. C
14. C
15. D
16. B
17. D
18. C
19. C
20. B
21. D
22. A
23. C
24. A
25. D
26. C
27. B
28. C
29. A
30. C
31. B
32. C
33. D
34. C
35. B
36. C
37. A
38. B
39. A
40. D
41. A
42. B
43. D
44. B
45. D
46. C
47. A
48. B
49. B
50. B

PREPARATION EXAM 4—EXPLANATIONS

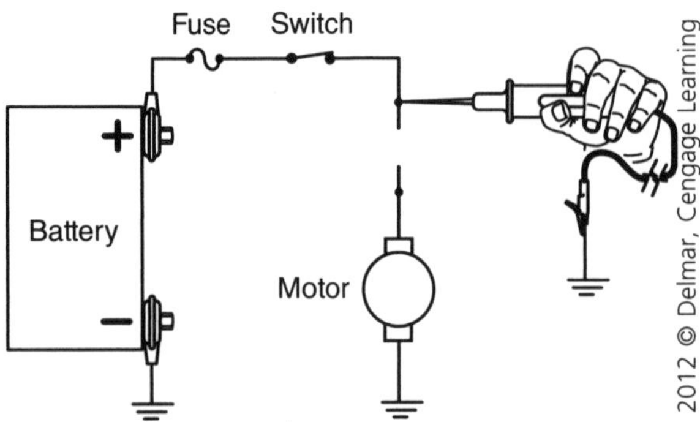

TASKS
A.8, D.8

1. Referring to the figure above, Technician A says that the test light will not light because there is an open circuit. Technician B says that the test light should not be used on this type of circuit. Who is correct?

   A. A only
   B. B only
   C. Both A and B
   D. Neither A nor B

Answer A is incorrect. The test light is connected to a point in the circuit that is before the open circuit so there is source voltage available to that point.

Answer B is incorrect. A test light can be used on circuits that do not have logic devices directly connected without any harm.

Answer C is incorrect. Neither Technician is correct.

Answer D is correct. Neither Technician is correct. The test light is connected before the open circuit so it should light up the light. It is acceptable to use test lights on motor circuits as long as the light is not connected to a data/logic circuit. A voltmeter will provide more exact feedback for the technician.

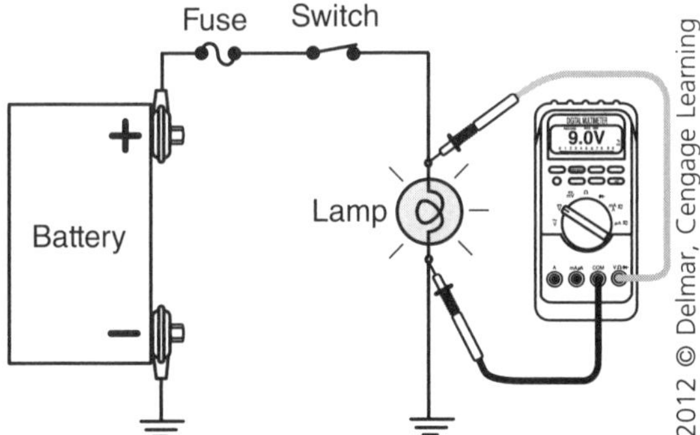

2. The circuit in the figure above is a 12 volt circuit and the battery is fully charged. Which of the following would be most likely to cause the reading?

TASK A.1

   A. Bad fuse
   B. Bad switch
   C. Bad lamp
   D. Open circuit at the switch

   Answer A is incorrect. A bad fuse would cause the meter to read zero volts.

   Answer B is correct. A switch with faulty contacts could cause the lamp to only receive nine volts.

   Answer C is incorrect. The lamp is only receiving nine volts. There is no evidence that the lamp is bad.

   Answer D is incorrect. An open circuit at the switch would cause the meter to read zero volts.

3. Which tool is recommended by manufacturers to perform voltage measurements on circuits that are controlled or monitored by a control module?

TASK A.1

   A. DMM
   B. Test light
   C. Continuity tester
   D. Oscilloscope

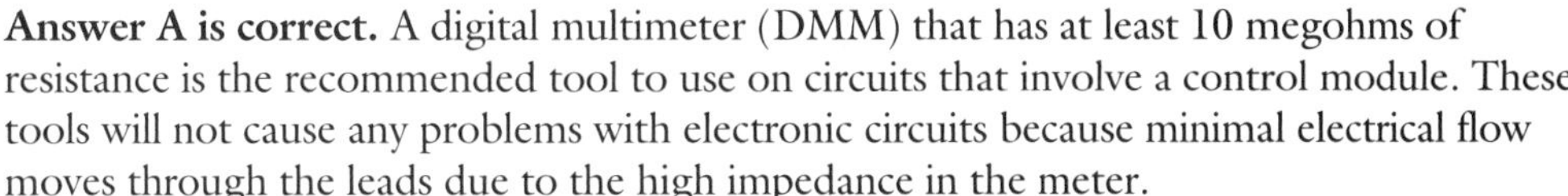

   Answer A is correct. A digital multimeter (DMM) that has at least 10 megohms of resistance is the recommended tool to use on circuits that involve a control module. These tools will not cause any problems with electronic circuits because minimal electrical flow moves through the leads due to the high impedance in the meter.

   Answer B is incorrect. A test light has very little resistance and could cause damage to a control module as well as cause a negative effect on the circuit being tested.

   Answer C is incorrect. A continuity tester is never used to check voltage in a circuit. It is only used on an un-powered circuit to check for a positive electrical connection (continuity).

   Answer D is incorrect. An oscilloscope is not required to perform a voltage measurement on a circuit. An oscilloscope is a useful tool that will show the voltage represented by a line on a screen (voltage over time).

TASK A.2

4. All of the following conditions will cause reduced current flow EXCEPT:

A. Loose ground connection
B. Shorted motor winding
C. Corroded terminal
D. Burnt connector

Answer A is incorrect. A loose ground connection would be a point of electrical resistance, which would reduce current flow.

Answer B is correct. A shorted motor winding would reduce the electrical resistance, which would cause the current flow to increase.

Answer C is incorrect. A corroded terminal would be a point of electrical resistance, which would reduce current flow.

Answer D is incorrect. A burnt connector would be a point of electrical resistance, which would reduce current flow.

TASK A.3

5. Technician A says that an open switch should have infinite resistance. Technician B says that a closed switch should have continuity. Who is correct?

A. A only
B. B only
C. Both A and B
D. Neither A nor B

Answer A is incorrect. Technician B is also correct.

Answer B is incorrect. Technician A is also correct.

Answer C is correct. Both Technicians are correct. An open switch would not provide any path for current to flow. An ohmmeter would measure infinite resistance when used to test an open switch. A closed switch does allow a path for current to flow. An ohmmeter would measure nearly zero ohms of resistance when used to test a closed switch. A continuity tester would show continuity when used to test a closed switch.

Answer D is incorrect. Both Technicians are correct.

TASK A.3

6. Technician A says that power must be turned off in the circuit before using an ohmmeter to make a measurement. Technician B says that the ohmmeter applies a small amount of voltage to the circuit to calculate resistance. Who is correct?

A. A only
B. B only
C. Both A and B
D. Neither A nor B

Answer A is incorrect. Technician B is also correct.

Answer B is incorrect. Technician A is also correct.

Answer C is correct. Both Technicians are correct. An ohmmeter should never be connected to a live electrical circuit because it produces its own voltage when configured as an ohmmeter. Hooking an ohmmeter to a live circuit will likely damage the meter.

Answer D in incorrect. Both Technicians are correct.

7. Viewing waveforms on an oscilloscope would be useful for all of the following conditions EXCEPT:

TASK A.4

   A. Checking for intermittent problems in a permanent magnet (PM) generator
   B. Checking a relay coil for a voltage spike
   C. Measuring the voltage drop across a motor
   D. Checking for excessive voltage spikes coming from the generator

   Answer A is incorrect. An intermittent fault with a PM generator could be viewed with an oscilloscope. The technician would notice a quick change in the waveform when the intermittent fault occurred.

   Answer B is incorrect. An oscilloscope would easily show the voltage spike from a relay coil when the coil is de-energized.

   Answer C is correct. An oscilloscope would not be the correct tool to use to measure the voltage drop in a circuit.

   Answer D is incorrect. An oscilloscope could be used to show voltage spikes coming from the generator.

8. Technician A says that a scan tool can retrieve codes from the body control module (BCM) on late model vehicles. Technician B says that a scan tool can display live data from some electronic systems. Who is correct?

TASK A.5

   A. A only
   B. B only
   C. Both A and B
   D. Neither A nor B

   Answer A is incorrect. Technician B is also correct.

   Answer B is incorrect. Technician A is also correct.

   Answer C is correct. Both Technicians are correct. Scan tools communicate with many of the electronic modules on cars and trucks. A technician can view live sensor and switch data as well as retrieve diagnostic trouble codes from many modules. A bi-directional scan tool has the ability to give commands to the vehicle and can be used to energize many of the outputs that the control modules control.

   Answer D is incorrect. Both Technicians are correct.

9. Technician A says that a short to ground before the load will cause the circuit protection device to open when the hot-side switch is turned on. Technician B says that a short to ground after the load will cause the circuit protection device to open when the hot-side switch is turned on. Who is correct?

TASK A.6

   A. A only
   B. B only
   C. Both A and B
   D. Neither A nor B

   Answer A is correct. Only Technician A is correct. This scenario would cause excessive current to flow because the electricity would take the path of least resistance and bypass the load. When this happens, the circuit protection device would heat up and open the circuit.

   Answer B is incorrect. A grounded circuit after the load would not negatively affect the circuit. This would just be a redundant ground located in the ground-side of the circuit.

   Answer C is incorrect. Only Technician A is correct.

   Answer D is incorrect. Technician A is correct.

TASK A.7

10. A vehicle is being diagnosed for a dead battery. During the diagnosis, an excessive parasitic draw is found. Technician A says that removing all of the fuses at once is an effective method of locating the unwanted draw. Technician B says that a "stuck open" trunk light switch could be the cause. Who is correct?

A. A only
B. B only
C. Both A and B
D. Neither A nor B

Answer A is incorrect. Removing all of the fuses at once would not show which circuit was causing the problem.

Answer B is incorrect. An open switch would not cause any lights to stay on.

Answer C is incorrect. Neither Technician is correct.

Answer D is correct. Neither Technician is correct. The correct method would be to remove the fuses one at a time while watching the ammeter. An open switch would not cause excessive parasitic draw.

TASK A.8

11. Technician A says that a fusible link will always burn into two pieces when it blows. Technician B says that a fusible link can be tested with a voltmeter without disconnecting it from the vehicle. Who is correct?

A. A only
B. B only
C. Both A and B
D. Neither A nor B

Answer A is incorrect. Some burnt fusible links do not burn in two pieces when they blow. The technician should test the voltage on each side of the link or simply pull on it to see if it pulls into two pieces.

Answer B is correct. Only Technician B is correct. A voltmeter can be used to test fusible links without disconnecting it from the circuit. A good fusible link should have source voltage on both sides of it.

Answer C is incorrect. Only Technician B is correct.

Answer D is incorrect. Technician B is correct.

TASK A.9

12. Which of the following details would be the most likely item to be located on a wiring diagram?

A. The power and ground distribution for the circuit
B. The location of the ground connection
C. Updated factory information about pattern failures
D. A flowchart for troubleshooting an electrical problem

Answer A is correct. A wiring diagram will provide details about how the power and ground are connected to the circuit.

Answer B is incorrect. A wiring diagram does not typically provide the location of electrical components such as a ground connection.

Answer C is incorrect. A wiring diagram does not typically provide updated factory information. Technical service bulletins provide updated factory information about pattern failures.

Answer D is incorrect. A wiring diagram does not provide any flowcharts for troubleshooting.

13. A wiring schematic is being used to troubleshoot an electrical problem. Technician A says that most schematics show the colors of the wires. Technician B says that most schematics are drawn with the power coming from the top of the picture. Who is correct?

TASK A.9

A. A only
B. B only
C. Both A and B
D. Neither A nor B

Answer A is incorrect. Technician B is also correct.

Answer B is incorrect. Technician A is also correct.

Answer C is correct. Both Technicians are correct. Most wiring schematic drawings have the wire colors labeled on the schematics. The colors are typically abbreviated with the letters in the color. For example R = red, W = white, G = green. If the wire has a tracer color, the schematic will be labeled R/W = red with a white tracer or G/W = green with a white tracer. As a general rule, the drafters of wire schematics draw the picture with the power coming from the top of the picture and ground coming from the bottom of the picture.

Answer D is incorrect. Both Technicians are correct.

14. Which tool is the LEAST LIKELY choice to use in the diagnosis of a problem on the data bus network?

TASK A.10

A. Oscilloscope
B. Digital multimeter
C. Continuity tester
D. Scan tool

Answer A is incorrect. An oscilloscope could be used to view electrical activity on the data bus network.

Answer B is incorrect. A digital multimeter could be used to measure voltage levels on the data bus network.

Answer C is correct. A continuity tester should not be used on data circuits on late model vehicles.

Answer D is incorrect. A scan tool could be used to retrieve data from the electronic modules that communicate on the data bus network.

15. Technician A says that a 12 volt battery that has 6 volts at the posts is 50 percent charged. Technician B says that a 12 volt battery that has 12.6 volts at the posts is overcharged. Who is correct?

TASK B.1

A. A only
B. B only
C. Both A and B
D. Neither A nor B

Answer A is incorrect. The 50 percent charge level on a 12 volt battery is 12.2 volts.

Answer B is incorrect. A battery with 12.6 volts at the posts is fully charged.

Answer C is incorrect. Neither Technician is correct.

Answer D is correct. Neither Technician is correct. A battery that has only 6 volts at the posts is severely discharged. A fully charged automotive battery should have 12.6 volts. These batteries have 6 cells that produce 2.1 volts each.

TASK B.2

16. A battery load test has been performed on an automotive battery. The voltage at the end of the 15 second test was 8.4 volts. Technician A says that the battery should be recharged for 15 minutes and then retested. Technician B says that the battery terminals sometimes get warm during this test. Who is correct?

A. A only
B. B only
C. Both A and B
D. Neither A nor B

Answer A is incorrect. The battery failed the load test and should be replaced. A good battery should have at least 9.6 volts at the end of a load test.

Answer B is correct. Only Technician B is correct. Load-testing batteries pulls large amounts of current from the battery and does cause the battery terminals to heat up during the test.

Answer C is incorrect. Only Technician B is correct.

Answer D is incorrect. Technician B is correct.

TASK B.4

17. Technician A says that overcharging a battery will not cause significant long-term damage. Technician B says that in hot weather, more current is needed to charge a battery. Who is right?

A. A only
B. B only
C. Both A and B
D. Neither A nor B

Answer A is incorrect. Overcharging will cause a battery to boil out electrolyte, and can also damage internal plates if it is severe enough.

Answer B is incorrect. A battery will more readily accept a charge in warmer weather, requiring less voltage to do so.

Answer C is Incorrect. Neither Technician is correct.

Answer D is correct. Neither Technician is correct. Overcharging the battery should be avoided due to possibly boiling the electrolyte out of the battery. Batteries can be charged with a lower charge level in hot weather.

TASK B.6

18. A vehicle is being jumpstarted. Technician A says the engine should be running on the boost vehicle before attempting to crank the dead vehicle. Technician B says the engine should be off while connecting the booster cables. Who is correct?

A. A only
B. B only
C. Both A and B
D. Neither A nor B

Answer A is incorrect. Technician B is also correct.

Answer B is incorrect. Technician A is also correct.

Answer C is correct. Both Technicians are correct. Having the engine running on boost vehicle is advisable because the charging system can charge the battery being boosted. It is a good practice to allow the boost vehicle to run 8–10 minutes on fast idle before attempting to start the dead vehicle. It is also a good practice to have the engine off while connecting the cables.

Answer D is incorrect. Both Technicians are correct.

19. A starter current draw test was performed on a late model vehicle. Technician A says that worn starter bushings will cause high current draw. Technician B says that battery terminal corrosion will cause lower than normal current draw. Who is correct?

TASK B.7

A. A only
B. B only
C. Both A and B
D. Neither A nor B

Answer A is incorrect. Technician B is also correct.

Answer B is incorrect. Technician A is also correct.

Answer C is correct. Both Technicians are correct. Worn starter bushings will cause increased physical resistance, which would cause more force to be needed to rotate the armature. This extra force would cause an increase in starter current draw. Battery corrosion would add electrical resistance, which would reduce starter current draw.

Answer D is incorrect. Both Technicians are correct.

20. A vehicle is being diagnosed for an inoperative starter. A voltage drop test was performed on the solenoid "load side" while the ignition switch is held in the crank mode and 0.2 volts measured. Technician A says that the solenoid is faulty. Technician B says that this test should only be performed if the battery is at least 75 percent charged. Who is correct?

TASK B.8

A. A only
B. B only
C. Both A and B
D. Neither A nor B

Answer A is incorrect. The voltage drop test revealed only 0.2 volts which is not excessive.

Answer B is correct. Only Technician B is correct. The vehicle battery should be at least 75 percent charged when testing the starter circuit.

Answer C is incorrect. Only Technician B is correct.

Answer D is incorrect. Technician B is correct.

21. A vehicle will not crank and the technician notices that the interior lights do not dim when the ignition switch is moved to the start position. The most likely cause would be which of the following?

TASKS B.9, D.6

A. Stuck closed starter relay
B. Loose battery cable connections
C. Starter mounting bolts loose
D. Stuck open ignition switch

Answer A is incorrect. A stuck closed starter relay would cause the starter to hang in the engaged position and would cause the starter motor to burn up.

Answer B is incorrect. Loose battery cable connections could cause a no-crank problem, but the dome lights would get dimmer or go out when the ignition switch was operated.

Answer C is incorrect. Loose starter mounting bolts could cause unusual starter sounds but would not typically cause a no-crank problem.

Answer D is correct. An ignition switch that is stuck open would cause this problem. This would cause no signal to be sent to the starter relay or solenoid, therefore the interior lights would not change as the ignition switch is operated.

TASK B.9

22. Which of the following functions would be LEAST LIKELY performed by the starter solenoid?

A. Prevents the armature from over-spinning
B. Push the drive gear out to the flywheel
C. Connects the "bat" terminal to the "motor" terminal
D. Provides a path for high current to flow

Answer A is correct. The starter solenoid has no control over the speed of the armature.

Answer B is incorrect. The starter solenoid creates linear movement to push the drive gear into the flywheel.

Answer C is incorrect. The starter solenoid acts as a switch to connect the "bat" terminal to the "motor" terminal when the starter is engaged.

Answer D is incorrect. The starter solenoid provides an electrical path for high current flow through the contacts and into the starter housing.

TASK B.9

23. Technician A says that the replacement starter assembly should be inspected carefully prior to installing on the engine. Technician B says that the replacement starter should be bench-tested prior to installing on the engine. Who is correct?

A. A only
B. B only
C. Both A and B
D. Neither A nor B

Answer A is incorrect. Technician B is also correct. At times damage can be incurred during shipping.

Answer B is incorrect. Technician A is also correct. There is always the possibility of defective components or misassembly.

Answer C is correct. Both Technicians are correct. All replacement starters should be carefully inspected as well as bench-tested prior to installing onto the engine.

Answer D is incorrect. Both Technicians are correct.

TASK C.1

24. A maintenance-free battery is low on electrolyte. Technician A says a defective voltage regulator may cause this problem. Technician B says a loose alternator belt may cause this problem. Who is right?

A. A only
B. B only
C. Both A and B
D. Neither A nor B

Answer A is correct. Only Technician A is correct. A defective voltage regulator can cause overcharging and possible battery boil over. Voltage regulators can be located in the alternator or in the engine control module depending on the application.

Answer B is incorrect. A loose alternator belt might cause undercharging but not overcharging, which is what low electrolyte level indicates.

Answer C is incorrect. Only Technician A is correct.

Answer D is incorrect. Technician A is correct.

25. Which of the following options would the most likely maximum charging voltage on a late-model vehicle?

TASK D.3

A. 13.4 volts
B. 15.8 volts
C. 15.6 volts
D. 14.6 volts

Answer A is incorrect. Many generators can create more than 13.4 volts maximum.

Answer B is incorrect. 15.8 volts is higher than most generators produce.

Answer C is incorrect. 15.6 volts is higher than most generators produce.

Answer D is correct. 14.6 volts is the approximate maximum charging voltage for most generators on late-model vehicles.

26. A late-model vehicle is being diagnosed for a charging problem. The generator only charges at 12.2 volts. Technician A says that the voltage drop should be checked on the charging output wire. Technician B says that the voltage drop should be checked on the charging ground circuit. Who is correct?

TASK C.6

A. A only
B. B only
C. Both A and B
D. Neither A nor B

Answer A is incorrect. Technician B is also correct.

Answer B is incorrect. Technician A is also correct.

Answer C is correct. Both Technicians are correct. Checking the voltage drop in the positive and negative side of the charging system is always recommended when a charging system is not performing up to specifications.

Answer D is incorrect. Both Technicians are correct.

27. During an output test of the charging system, a technician finds that the charging current is at 75 amps. The specification for the vehicle is 115 amps. What is the most likely cause?

TASK C.4

A. The vehicle battery has the wrong CCA.
B. The output test was performed at idle.
C. The battery cables are oversize.
D. The voltage regulator has stuck at 100 percent.

Answer A is incorrect. Having the wrong battery can cause electrical problems such as slow cranking or dim lights at idle with high electrical loads, but it should not cause the charging current to be low during an output test.

Answer B is correct. The RPM needs to be increased to 1500–2000 during a charging output test in order for the technician to get accurate results.

Answer C is incorrect. Having oversized battery cables would allow current to flow easier to the electrical system of the vehicle.

Answer D is incorrect. A voltage regulator stuck at 100 percent would cause an over-charging problem. If this happens, the technician would likely notice the battery heating up and creating a terrible odor under the hood.

TASKS
A.5, B.5

28. A PCM controlled charging system is being diagnosed for a low charging problem. Technician A says that a bi-directional scan tool can be used to test the voltage regulation function for fault codes. Technician B says that the alternator can be commanded to charge at full capacity with a bi-directional scan tool. Who is correct?

A. A only
B. B only
C. Both A and B
D. Neither A nor B

Answer A is incorrect. Technician B is also correct.

Answer B is incorrect. Technician A is also correct.

Answer C is correct. Alternators that are PCM controlled can be diagnosed for fault codes with a bi-directional scan tool. The alternators can also be commanded to charge at full capacity by using the "output test" function of the scan tool.

Answer D is incorrect. Both technicians are correct.

TASKS
D.1, D.4, D.6

29. An inoperative lamp is being diagnosed. Technician A says that an open wire between the switch and the bulb could be the cause. Technician B says that a shorted switch could be the cause. Who is correct?

A. A only
B. B only
C. Both A and B
D. Neither A nor B

Answer A is correct. Only Technician A is correct. An open wire can cause electrical loads to be inoperative. Current does not flow through an open circuit.

Answer B is incorrect. A shorted switch would cause the circuit to stay on all of the time. It would not cause the inoperative lamp.

Answer C is incorrect. Only Technician A is correct.

Answer D is incorrect. Technician A is correct.

TASKS
D.2, D.3

30. The headlights work on high beams but are inoperative on low beams. Technician A says that the dimmer switch could be the fault. Technician B says that both bulbs could have open low beam filaments. Who is correct?

A. A only
B. B only
C. Both A and B
D. Neither A nor B

Answer A is incorrect. Technician B is also correct.

Answer B is incorrect. Technician A is also correct.

Answer C is correct. Both Technicians are correct. A faulty dimmer switch could cause the headlights to be inoperative on low beams and possibly still work on high beams. Though unlikely, it is possible that both headlights could be blown on the low beam filaments. A quick voltage test at the headlight assembly would produce the correct diagnosis. If the bulbs are receiving system voltage with the dimmer switch on low beams, then the bulbs would be at fault.

Answer D is incorrect. Both Technicians are correct.

31. The trunk light stays on continuously on a late model vehicle. Technician A says that an open trunk light switch is the likely cause. Technician B says that a shorted wire near the trunk light switch could be the cause. Who is correct?

TASK D.6

A. A only
B. B only
C. Both A and B
D. Neither A nor B

Answer A is incorrect. An open trunk light switch would not cause the trunk light to stay on. The trunk switch turns on the trunk light by closing its contacts.

Answer B is correct. Only Technician B is correct. A shorted wire near the trunk light switch could cause the trunk light to stay on continuously.

Answer C is incorrect. Only Technician B is correct.

Answer D is incorrect. Technician B is correct.

32. Technician A says that many brake light switches are adjustable. Technician B says that many brake light switches also have circuits that disengage the cruise control. Who is correct?

TASKS D.4, E.4, F.9

A. A only
B. B only
C. Both A and B
D. Neither A nor B

Answer A is incorrect. Technician B is also correct.

Answer B is incorrect. Technician A is also correct.

Answer C is correct. Both Technicians are correct. Many brake light switches are adjustable. In addition, brake light switches also have circuits that disengage the cruise control when the brake pedal is depressed.

Answer D is incorrect. Both Technicians are correct.

33. Technician A says that the turn signal flasher is not load sensitive. Technician B says that the hazard light flasher is load sensitive. Who is correct?

TASK D.4

A. A only
B. B only
C. Both A and B
D. Neither A nor B

Answer A is incorrect. Most turn signal flashers change their flash rate when a bulb burns out.

Answer B is incorrect. Most hazard flashers are not load sensitive. These devices flash the signal lamps at all corners of the vehicle.

Answer C is incorrect. Neither Technician is correct.

Answer D is correct. Neither Technician is correct. Turn signal flashers are usually sensitive to electrical load. They are designed this way in order to provide feedback to the driver when a bulb burns out. Hazard flashers are not typically sensitive to electrical load.

TASKS
D.4, F.9

34. The bulbs for the back-up lights have both failed. Technician A installs some dielectric grease on the terminals of the replacement bulb before installing. Technician B tests the voltage available to the back-up lights after replacing the bulbs. Who is correct?

   A. A only
   B. B only
   C. Both A and B
   D. Neither A nor B

Answer A is incorrect. Technician B is also correct.

Answer B is incorrect. Technician A is also correct.

Answer C is correct. Both Technicians are correct. It is advisable to put some dielectric grease on all bulb contacts before installing to help the bulb and socket resist water intrusion. Testing the voltage to the backup bulbs would also be a good practice since both bulbs have failed.

Answer D is incorrect. Both Technicians are correct.

TASK E.5

35. The temperature sending unit needs to be replaced on a late-model vehicle. Technician A says that the whole cooling system will need to be drained during this process. Technician B says that the engine should be cooled down prior to performing this repair. Who is correct?

   A. A only
   B. B only
   C. Both A and B
   D. Neither A nor B

Answer A is incorrect. Draining the whole cooling system is not required when replacing the temperature sending unit.

Answer B is correct. Only Technician B is correct. The engine should be cooled down to be safer for the technician performing the repair.

Answer C is incorrect. Only Technician B is correct.

Answer D is incorrect. Technician B is correct.

TASKS
E.3, E.5

36. Technician A says that a corroded connection at the fuel sending unit could cause a electromagnetic fuel gauge to read too high. Technician B says that a grounded connection at the fuel sending unit could cause a bi-metallic fuel gauge to read high. Who is correct?

   A. A only
   B. B only
   C. Both A and B
   D. Neither A nor B

Answer A is incorrect. Technician B is also correct.

Answer B is incorrect. Technician A is also correct.

Answer C is correct. Both Technicians are correct. A corroded connection would increase the resistance in the fuel gauge circuit, which would cause the electromagnetic gauge to read higher. A grounded connection would reduce the electrical resistance, which would cause the bi-metallic gauge to read higher.

Answer D is incorrect. Both Technicians are correct.

37. The fuel gauge reads full at all times. All of the other instrument panel gauges work correctly. Technician A says that the fuel sending unit wire could be broken. Technician B says that the gauges fuse could be blown. Who is correct?

TASKS E.3, E.5

A. A only
B. B only
C. Both A and B
D. Neither A nor B

Answer A is correct. Only Technician A is correct. A broken fuel sending unit wire could cause the fuel gauge to read full at all times on an electromagnetic gauge.

Answer B is incorrect. All of the gauges would be inoperative if the gauges fuse were blown.

Answer C is incorrect. Only Technician A is correct.

Answer D is incorrect. Technician A is correct.

38. Warning lights and warning devices are generally activated by which of the following components?

TASK E.4

A. An ISO relay
B. Closing of a switch or sensor
C. Opening of a switch or sensor
D. An inertia switch

Answer A is wrong. ISO relays have many uses on cars and trucks, but warning lights and devices typically do not use them to turn on warning lights.

Answer B is correct. Closing a switch or sensor is usually required to complete a circuit to power-up a light or warning device.

Answer C is wrong. Opening a switch or sensor will interrupt current flow in a circuit, canceling the operation of a light or warning device.

Answer D is wrong. Inertia switches are not typically used to activate warning lights and devices. These switches are commonly used in inflatable restraint systems to detect rapid deceleration.

39. The indicator for the bright lights does not work but the bright and dim headlights function correctly. Technician A says that the bulb for the bright indicator could be defective. Technician B says that the dimmer switch could be defective. Who is correct?

TASKS D.2, E.4

A. A only
B. B only
C. Both A and B
D. Neither A nor B

Answer A is correct. Only Technician A is correct. The bulb for the bright indicator could be the cause of this problem. The technician could inspect the wiring diagram to find a test point to measure voltage going to the indicator bulb.

Answer B is incorrect. The dimmer switch is working because the headlights function on both bright and dim.

Answer C is incorrect. Only Technician A is correct.

Answer D is incorrect. Technician A is correct.

TASK E.4

40. The chime reminder for the headlights sounds at times as the vehicle is driven over rough roads. Which of the following conditions would be the most likely cause for this problem?

A. Open wire in the headlight circuit
B. Faulty headlight bulb
C. Park light bulb burnt out
D. Headlight wire rubbing an instrument panel bracket

Answer A is incorrect. An open wire in the headlight circuit would not cause this intermittent chime problem.

Answer B is incorrect. The chime reminder would not sound when driving over rough roads if the headlight bulb was faulty.

Answer C is incorrect. The chime reminder would not sound when driving over rough roads if the park light was burned out.

Answer D is correct. A headlight wire rubbing a panel bracket could be causing this intermittent problem.

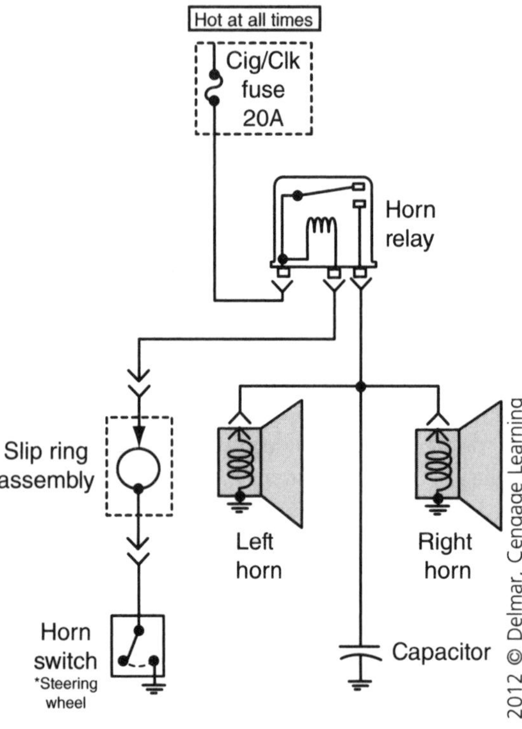

TASK F.10

41. Referring to the figure above, Technician A says that the "Cig/Clk" fuse powers both sides of the horn relay. Technician B says that horn switch provides voltage to the relay coil when the switch is depressed. Who is correct?

A. A only
B. B only
C. Both A and B
D. Neither A nor B

Answer A is correct. Only Technician A is correct. The "Cig/Clk" fuse supplies voltage to the coil side and the load side of the horn relay.

Answer B is incorrect. The horn switch provides a ground for the relay coil when the switch is depressed.

Answer C is incorrect. Only Technician A is correct.

Answer D is incorrect. Technician A is correct.

42. What is the most likely cause of a windshield wiper system that only works on low speed?

TASK F.9

A. A blown fuse

B. A faulty multi-function switch

C. A lose ground at the wiper motor

D. An open park switch

Answer A is incorrect. A blown fuse would cause the wipers to be totally inoperative on all speeds.

Answer B is correct. The multi-function switch often contains the wiper switch in addition to several other switches, such as the turn signal switch, the hazard switch, the headlight switch, and sometimes the cruise control switch.

Answer C is incorrect. A loose wiper motor ground would cause all of the speeds to be negatively affected.

Answer D is incorrect. An open park switch on the windshield wiper motor would cause the wipers to fail to return to the rest/park position when the switch is turned off.

43. The windshield washer system does not spray any water on the windshield when the switch is depressed. A voltage test was performed at the pump motor while the switch was depressed and 13 volts was measured. What is the most likely cause of this problem?

TASK F.10

A. A faulty washer pump switch

B. A faulty washer pump relay

C. An open wire in the washer pump circuit

D. A faulty washer pump motor

Answer A is incorrect. The pump motor is dropping 13 volts so the switch is good.

Answer B is incorrect. The pump motor would be totally inoperative and not drop any voltage if the relay was faulty.

Answer C is incorrect. The pump motor would be totally inoperative and not drop any voltage if there was an open wire.

Answer D is correct. The voltage test revealed that 13 volts was being measured at the pump motor with the switch depressed. This proves that the motor has a good power and ground connection, so the motor is the problem.

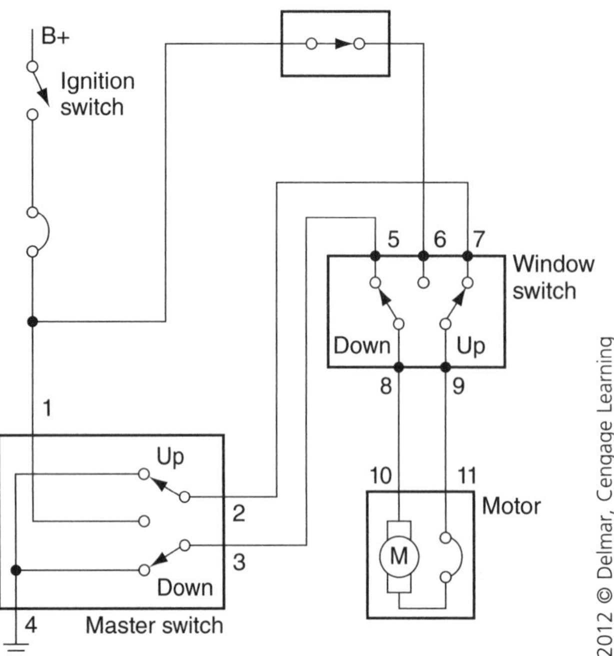

TASK F.1

44. Referring to the figure above, the power window motor operates correctly from the master switch but does not operate at all from the window switch. Technician A says that the built-in circuit breaker in the motor could be defective. Technician B says that terminal #6 at the window switch could be broken. Who is correct?

A. A only
B. B only
C. Both A and B
D. Neither A nor B

Answer A is incorrect. The motor would not operate from any switch if the circuit breaker was defective.

Answer B is correct. Only Technician B is correct. A broken terminal #6 would cause the window switch to not receive battery voltage to be able to operate independently from the master switch.

Answer C is incorrect. Only Technician B is correct.

Answer D is incorrect. Technician B is correct.

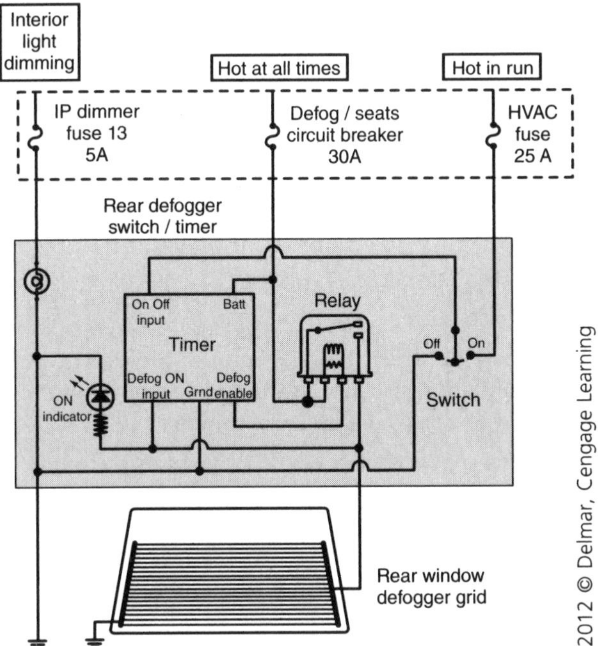

45. Referring to the figure above, Technician A says that the HVAC fuse supplies the "load side" of the defogger relay. Technician B says that the HVAC fuse is hot at all times. Who is correct?

TASK F.3

A. A only

B. B only

C. Both A and B

D. Neither A nor B

Answer A is incorrect. The "defog/seats" circuit breaker supplies the "load side" of the defogger relay.

Answer B is incorrect. The HVAC fuse is hot in run.

Answer C is incorrect. Neither Technician is correct.

Answer D is correct. Neither Technician is correct. The HVAC fuse is only hot in run. The HVAC fuse supplies power to the rear defogger switch.

TASK F.2

46. An inoperative sunroof is being diagnosed. A blown fuse for the sunroof is found. Technician A says the cables should be checked for a binding problem. Technician B says that the motor should be checked for a shorting problem with an ohmmeter. Who is correct?

A. A only
B. B only
C. Both A and B
D. Neither A nor B

Answer A is incorrect. Technician B is also correct.

Answer B is incorrect. Technician A is also correct.

Answer C is correct. Both Technicians are correct. Binding sunroof cables could cause the circuit amperage to suddenly increase and blow the fuse. A shorted sunroof motor could also cause a blown fuse. This problem could be checked with an ohmmeter.

Answer D is incorrect. Both Technicians are correct.

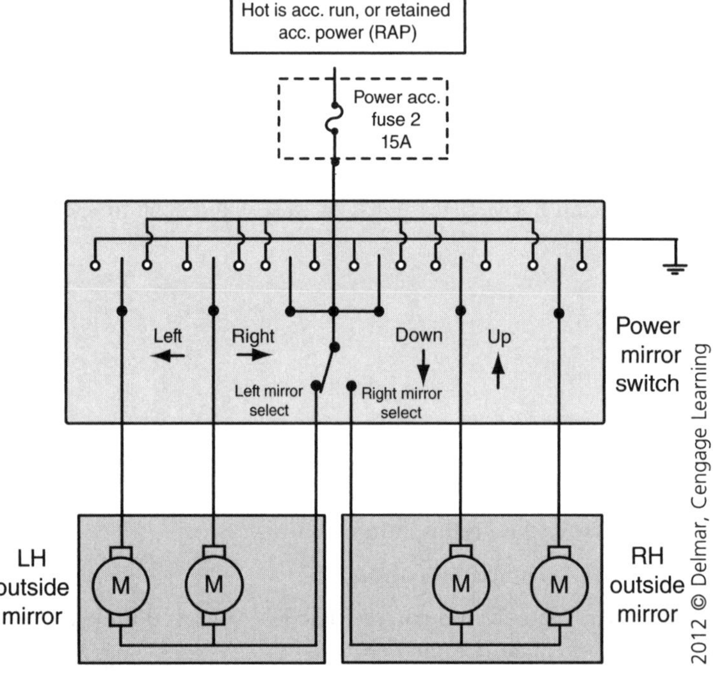

TASK F.1

47. Referring to the figure above, the left mirror functions normally but the right side power mirror does not function at all. Technician A says that mirror select switch could be defective. Technician B says that the power mirror switch could have a disconnected ground. Who is correct?

A. A only
B. B only
C. Both A and B
D. Neither A nor B

Answer A is correct. Only Technician A is correct. A defective mirror select switch could cause the right side mirror to not operate at all.

Answer B is incorrect. A disconnected ground on the power mirror switch would cause all power mirror operation to cease.

Answer C is incorrect. Only Technician A is correct.

Answer D is incorrect. Technician A is correct.

48. A radio assembly is being replaced on a late-model vehicle. Technician A says that the radio fuse should be removed for 30 minutes prior to removing the radio. Technician B says that the anti-theft unlock code has to be programmed into the radio before it will function. Who is correct?

TASKS F.6, F.8

A. A only

B. B only

C. Both A and B

D. Neither A nor B

Answer A is incorrect. There is no reason to remove the radio fuse prior to removing the radio.

Answer B is correct. Only Technician B is correct. The radio unlock code must be programmed into many late-model vehicles any time the radio loses battery power.

Answer C is incorrect. Only Technician B is correct.

Answer D is incorrect. Technician B is correct.

49. A late-model vehicle is being diagnosed with an anti-theft problem. Technician A says that the system can usually be overridden by using a jumper wire at the data link connector (DLC). Technician B says that a scan tool can be used to retrieve fault codes from the anti-theft system. Who is correct?

TASKS A.5, F.9

A. A only

B. B only

C. Both A and B

D. Neither A nor B

Answer A is incorrect. Installing a jumper wire at the data link connector (DLC) will not override the anti-theft system.

Answer B is correct. Only Technician B is correct. Scan tools can be used to retrieve fault codes from anti-theft systems.

Answer C is incorrect. Only Technician B is correct.

Answer D is incorrect. Technician B is correct.

50. A digital clock loses the correct time each time the ignition is turned off. The most likely cause of this problem is:

TASKS F.1, F.7

A. Faulty body control module (BCM)

B. Blown "keep alive memory" fuse

C. Faulty gateway module

D. Blown radio fuse

Answer A is incorrect. A fault in the BCM would likely cause additional problems beyond the digital clock losing the correct time.

Answer B is correct. The digital clock needs to have constant battery power to maintain the correct time. If the clock loses this constant feed, then it will lose the correct setting each time the ignition is shut off.

Answer C is incorrect. A faulty gateway module will greatly impact many of the electronic functions of the vehicle. It is not likely to just cause a problem with the clock losing the correct time.

Answer D is incorrect. A blown radio fuse will cause the radio to be totally inoperative and not just cause the clock to lose the correct setting.

PREPARATION EXAM 5—ANSWER KEY

1. A
2. C
3. B
4. D
5. A
6. C
7. D
8. B
9. C
10. B
11. B
12. B
13. B
14. D
15. B
16. B
17. D
18. C
19. D
20. D
21. C
22. A
23. B
24. B
25. C
26. A
27. D
28. B
29. C
30. C
31. B
32. A
33. D
34. A
35. A
36. B
37. C
38. D
39. B
40. C
41. D
42. C
43. B
44. A
45. C
46. B
47. C
48. B
49. A
50. C

PREPARATION EXAM 5—EXPLANATIONS

TASK F.3

1. A late-model vehicle is being diagnosed for an inoperative heated steering wheel. Technician A says that a faulty clock spring could cause this problem. Technician B says that the steering wheel heater grid can be replaced without removing the steering wheel. Who is correct?

   A. A only
   B. B only
   C. Both A and B
   D. Neither A nor B

   Answer A is correct. Only Technician A is correct. A clock spring with an open circuit could prevent the heated steering wheel from receiving power.

   Answer B is incorrect. The heater grid is built into the steering wheel and is not typically serviceable. The whole steering wheel must be replaced if the heater grid is faulty.

   Answer C is incorrect. Only Technician A is correct.

   Answer D is incorrect. Technician A is correct.

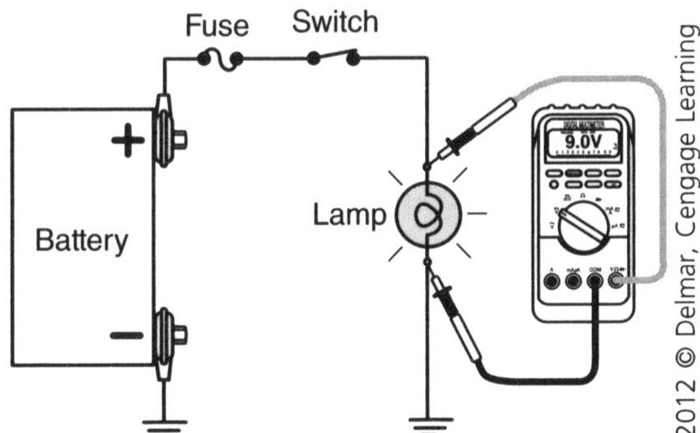

2. The circuit in the figure above is a 12 volt circuit and the battery is fully charged. Technician A says that the switch could have burnt contacts. Technician B says that the circuit could have a poor ground connection. Who is correct?

TASK A.1

A. A only

B. B only

C. Both A and B

D. Neither A nor B

Answer A is incorrect. Technician B is also correct.

Answer B is incorrect. Technician A is also correct.

Answer C is correct. Both Technicians are correct. The lamp is only dropping 9 volts so there is a problem somewhere in the circuit. A switch with burnt contacts could be the cause of the problem. A poor ground connection could also cause the problem. To further diagnose the problem, the technician would need to perform voltage drop tests on the switch and the ground to see where the voltage is being dropped.

Answer D is incorrect. Both Technicians are correct.

3. Technician A says that high current draw can be caused by dirty electrical connections. Technician B says that low current draw can be caused by corroded electrical connections. Who is correct?

TASK A.2

A. A only

B. B only

C. Both A and B

D. Neither A nor B

Answer A is incorrect. Dirty electrical connections will cause reduced current draw because of added electrical resistance.

Answer B is correct. Only Technician B is correct. Corroded connections will cause reduced current to flow because of the added electrical resistance.

Answer C is incorrect. Only Technician B is correct.

Answer D is incorrect. Technician B is correct.

TASK A.3

4. Technician A says that burnt electrical contacts will decrease the electrical resistance in a circuit. Technician B says that an open switch should have continuity. Who is correct?

   A. A only
   B. B only
   C. Both A and B
   D. Neither A nor B

Answer A is incorrect. Burnt electrical contacts will increase electrical resistance in a circuit.

Answer B is incorrect. An open switch should measure infinite resistance.

Answer C is incorrect. Neither Technician is correct.

Answer D is correct. Neither Technician is correct. Burnt electrical contacts will increase the resistance in a circuit, which will cause reduced current flow. An open switch will have infinite resistance. A closed switch should have continuity.

TASK A.4

5. Which of the following tests would be most likely performed with an oscilloscope?

   A. Inspecting a permanent magnet (PM) generator for a chipped tooth on the reluctor ring
   B. Testing the voltage drop in the positive battery cable
   C. Inspecting a relay coil resistance
   D. Testing the voltage drop on the charging output circuit

Answer A is correct. This problem could be located with an oscilloscope by carefully watching the waveform. The pattern would quickly change when the chipped tooth passed by the PM generator.

Answer B is incorrect. A digital voltmeter would be needed to test the voltage drop in a battery cable.

Answer C is incorrect. A digital ohmmeter would be needed to test the relay coil resistance.

Answer D is incorrect. A digital voltmeter would be needed to test the voltage drop on the charging output circuit.

TASK A.5

6. Technician A says that a scan tool can retrieve diagnostic trouble codes from the airbag control module on late model vehicles. Technician B says that a scan tool can display live data from the body control module. Who is correct?

   A. A only
   B. B only
   C. Both A and B
   D. Neither A nor B

Answer A is incorrect. Technician B is also correct.

Answer B is incorrect. Technician A is also correct.

Answer C is correct. Both Technicians are correct. Scan tools can be used to retrieve live data and diagnostic trouble codes from most of the control modules used on the vehicle. Some scan tools are bi-directional which allows the technician to perform functional tests on some of the outputs controlled by the various modules.

Answer D is incorrect. Both Technicians are correct.

7. A fused jumper wire can be used for all of the following procedures EXCEPT:

TASKS A.6, D.1, F.4

A. Bypassing the dimmer switch during diagnosis
B. Bypassing the rear defogger relay during diagnosis
C. Bypassing the blower motor relay during diagnosis
D. Bypassing a potentiometer reference voltage to ground during diagnosis

Answer A is incorrect. A fused jumper wire could be used to bypass the dimmer switch.

Answer B is incorrect. A fused jumper wire could be used to bypass the defogger relay.

Answer C is incorrect. A fused jumper wire could be used to bypass the blower motor relay.

Answer D is correct. Using a fused jumper to bypass the reference voltage of a potentiometer is not recommended. Doing this would likely cause control module problems.

8. Technician A says that a circuit with corrosion in the wiring will create a short circuit. Technician B says that a power wire that rubs a metal surface for a long period of time could create a short to ground. Who is correct?

TASK A.6

A. A only
B. B only
C. Both A and B
D. Neither A nor B

Answer A is incorrect. Corrosion in the wiring will not cause a short circuit.

Answer B is correct. Only Technician B is correct. A power wire that rubs a metal surface could cause a short to ground. A circuit protection device would likely open if a short to ground occurs.

Answer C is incorrect. Only Technician B is correct.

Answer D is incorrect. Technician B is correct.

9. Technician A says that an amp clamp is a useful tool to use when checking for key-off drain. Technician B says that a faulty rectifier bridge in the alternator could cause excessive key-off drain. Who is correct?

TASK A.8

A. A only
B. B only
C. Both A and B
D. Neither A nor B

Answer A is incorrect. Technician B is also correct.

Answer B is incorrect. Technician A is also correct.

Answer C is correct. Both Technicians are correct. An amp clamp allows a technician to install the tool around the positive or negative battery cables and accurately measure current flow. The advantage to using this tool is that the battery does not have to be disconnected to install the ammeter. If a diode shorts out in the rectifier bridge in the alternator then electrical current can flow from the battery into the alternator and cause the battery to run down.

Answer D is incorrect. Both Technicians are correct.

TASK A.8

10. Technician A says that circuit breakers are solid-state components. Technician B says that some circuit breakers automatically reset when they cool down. Who is correct?

A. A only

B. B only

C. Both A and B

D. Neither A nor B

Answer A is incorrect. Circuit breakers have a moveable contact so they are not considered to be solid state.

Answer B is correct. Only Technician B is correct. Some circuit breakers will reset automatically when they cool down. Other types of circuit breakers have to be manually reset after tripping.

Answer C is incorrect. Only Technician B is correct.

Answer D is incorrect. Technician B is correct.

TASK A.9

11. Technician A says that wiring diagrams give the location of each splice used in the circuit. Technician B says that wiring diagrams use electrical symbols to represent the electrical components. Who is correct?

A. A only

B. B only

C. Both A and B

D. Neither A nor B

Answer A is incorrect. Wiring diagrams do not usually give the location of the electrical components shown.

Answer B is correct. Only Technician B is correct. Wiring diagrams use symbols to represent the electrical components. Most of the symbols are universal and are used by many different manufacturers.

Answer C is incorrect. Only Technician B is correct.

Answer D is incorrect. Technician B is correct.

TASK A.10

12. A vehicle with a data bus network problem is being diagnosed. Technician A says that the owner will not likely notice any unusual problems if the data bus wires become shorted together. Technician B says data bus communication happens when voltage pulses are sent from module to module thousands of times per second. Who is correct?

A. A only

B. B only

C. Both A and B

D. Neither A nor B

Answer A is incorrect. The owner would likely notice problems in several systems including the engine, transmission, instrument cluster and climate control.

Answer B is correct. Only Technician B is correct. Data bus communication takes place over one or two wires by pulsing voltage at a very fast rate. Some networks can communicate up to 250,000 times per second. This rate is getting faster as vehicles become more and more complicated and high tech.

Answer C is incorrect. Only Technician B is correct.

Answer D is incorrect. Technician B is correct.

13. All of the following methods are possible methods of determining the state-of-charge of an automotive battery EXCEPT:

TASK E.4

A. Testing the specific gravity
B. Load test
C. Open circuit voltage test
D. Digital battery test

Answer A is incorrect. Testing the specific gravity is a way to determine the charge level on a battery that has removable caps.

Answer B is correct. A load test will not determine the state-of-charge. This test measures the battery's capacity to produce current for a period of time. The battery needs to be at least 75 percent charged in order for this test to be valid.

Answer C is incorrect. Testing the open circuit voltage is an accurate way to determine the charge level of a battery. A fully charged battery should have about 12.6 volts.

Answer D is incorrect. A digital battery tester provides the state-of-charge of an automotive battery. If the battery is not charged up, the tester will provide results that it needs to be charged and then retested.

14. A digital battery tester would be most likely used for which of the following purposes?

TASK E.4

A. Measuring resistance of control modules
B. Testing starter current draw
C. Measuring voltage of control modules
D. Testing the impedance of the battery

Answer A is incorrect. A technician should never attempt to measure the resistance of a control module. A digital ohmmeter could be used to measure resistance electronic sensors and circuits after they are disconnected from the circuit.

Answer B is incorrect. An electronic tool such as a DMM used with an amp clamp or possibly an electrical system tester that utilizes an amp clamp can be used to test starter draw.

Answer C is incorrect. A high-impedance DMM set to read DC volts could be used to read voltage levels of control module circuits.

Answer D is correct. A digital battery tester can be used to test the impedance of a vehicle battery. It is the recommended tool in the industry because it can accurately test a battery even if it just needs to be charged. Test results from this tester are very conclusive on what needs to be done with the battery.

TASK B.2

15. A battery load test has been performed on an automotive battery. The voltage at the end of the 15 second test was 10.4 volts. Technician A says that the battery should be recharged for 15 minutes and then retested. Technician B says that the connectors of the test leads sometimes get warm during this test. Who is correct?

A. A only

B. B only

C. Both A and B

D. Neither A nor B

Answer A is incorrect. The battery does not need to be recharged. If a battery has at least 9.6 volts at the end of this test, then it passes and can be put back in service.

Answer B is correct. Only Technician B is correct. It is not uncommon for the test lead connectors to get warm during this test.

Answer C is incorrect. Only Technician B is correct.

Answer D is incorrect. Technician B is correct.

TASK B.5

16. The battery housing received some damage from driving a vehicle on rough roads. Electrolyte spilled all over the battery tray. Technician A says that brake cleaner should be used to clean the area. Technician B says that baking soda could be used to neutralize the battery acid. Who is correct?

A. A only

B. B only

C. Both A and B

D. Neither A nor B

Answer A is incorrect. Brake cleaner should never be used around spilled electrolyte.

Answer B is correct. Only Technician B is correct. Baking soda can be used to neutralize the spilled electrolyte acid of a battery.

Answer C is incorrect. Only Technician B is correct.

Answer D is incorrect. Technician B is correct.

TASK B.7

17. A starter current draw test was performed on a late model vehicle. Technician A says that worn starter brushes will cause high current draw. Technician B says that battery terminal corrosion will cause higher than normal current draw. Who is correct?

A. A only

B. B only

C. Both A and B

D. Neither A nor B

Answer A is incorrect. Worn starter brushes would be a point of electrical resistance, which would cause lower current draw.

Answer B is incorrect. Battery terminal corrosion would be a point of electrical resistance, which would cause lower current draw.

Answer C is incorrect. Neither Technician is correct.

Answer D is correct. Neither Technician is correct. Both worn brushes and corroded battery cables will cause lower than normal starter current draw.

18. A vehicle is being diagnosed for an inoperative starter. A voltage drop test was performed on the solenoid "load side" while ignition switch is held in the crank mode, and 3 volts is measured. Technician A says that the solenoid is faulty. Technician B says that this test should only be performed if the battery is at least 75 percent charged. Who is correct?

TASK B.8

A. A only

B. B only

C. Both A and B

D. Neither A nor B

Answer A is incorrect. Technician B is also correct.

Answer B is incorrect. Technician A is also correct.

Answer C is correct. Both Technicians are correct. A 3 volt voltage drop on the "load side" of a solenoid indicates that the solenoid is faulty. The battery should be at least 75 percent charge level for this test to be valid.

Answer D is incorrect. Both Technicians are correct.

19. The starter solenoid performs all of the following functions EXCEPT:

TASK B.9

A. Provides a path for high current to flow into the starter

B. Push the drive gear out to the flywheel

C. Connects the "bat" terminal to the "motor" terminal

D. Provides gear reduction to increase torque in the starter

Answer A is incorrect. The starter solenoid provides a path for high current to flow into the starter.

Answer B is incorrect. The starter solenoid provides the linear movement to push the drive gear into the flywheel.

Answer C is incorrect. The starter solenoid connects the "bat" terminal to the "motor" terminal when it is energized.

Answer D is correct. The starter solenoid does not provide any gear reduction for the starter.

20. All of these conditions would cause the starter not to crank the engine **EXCEPT**:

TASK B.10

A. The battery is not connected to the starter motor.

B. The solenoid does not engage the starter drive pinion with the engine flywheel.

C. Failure of the control circuit to switch the large-current circuit.

D. The starter drive pinion fails to disengage from the flywheel.

Answer A is incorrect. If the battery cable was not connected to the starter motor, the starter would not rotate.

Answer B is incorrect. If the solenoid does not engage the pinion with the engine flywheel, the engine will not crank.

Answer C is incorrect. The control circuit is supposed to switch the large current circuit in order to crank the engine. This is the function of the starter solenoid and the magnetic switch.

Answer D is correct. A drive pinion that fails to disengage will not cause an engine to not start. When the engine starts, the flywheel spins the pinion faster than the armature. This action releases the rollers, unlocking the pinion gear from the armature shaft. The pinion then "overruns" the armature shaft freely until being pulled out of the mesh without stressing the starter motor. Note that the overrunning clutch is moved in and out of mesh with the flywheel by linkage operated by the solenoid.

TASK C.1

21. All of the following conditions could cause the generator to have zero output EXCEPT:

A. Blown fusible link in charge circuit
B. An open rotor coil
C. A full-fielded rotor
D. A faulty voltage regulator

Answer A is incorrect. A blown fusible link in the charge circuit would cause the generator to have zero output.

Answer B is incorrect. An open rotor coil would cause the generator to have zero output.

Answer C is correct. A full-fielded rotor would cause the generator to charge at maximum capacity.

Answer D is incorrect. A faulty voltage regulator could cause the generator to have zero output.

TASK C.2

22. An alternator with a 120-ampere rating produces 85 amps during an output test. A V-belt drives the alternator and the belt is at the specified tension. Technician A says the V-belt may be worn and bottomed in the pulley. Technician B says the alternator pulley may be misaligned with the crankshaft pulley. Who is right?

A. A only
B. B only
C. Both A and B
D. Neither A nor B

Answer A is correct. Only Technician A is correct. Even though the belt is at the specified tension, it may slip because it is bottomed in the pulley. If a belt tension gauge is not available, belt tension can be determined by depressing the belt at the center of its span. As a rule, 3/8", or 9.53 mm, is the approximate distance that the belt should be allowed to move.

Answer B is incorrect. A misaligned pulley should not cause low alternator output because of belt slippage. It may, however, cause the belt to jump off, in which case there would be zero output, or it may cause rapid belt wear.

Answer C is incorrect. Only Technician A is correct.

Answer D is incorrect. Technician A is correct.

TASK C.3

23. Which of the following options would be the LEAST LIKELY maximum charging voltage on a late model vehicle?

A. 13. 9 volts
B. 15.8 volts
C. 14. 3 volts
D. 14.6 volts

Answer A is incorrect. 13.9 volts is a possible peak charge voltage for an automotive generator.

Answer B is correct. 15.8 volts is too high of a peak charge voltage for an automotive generator. A charging voltage of this level could cause problems with the control modules on the vehicle.

Answer C is incorrect. 14.3 volts is a possible peak charge voltage for an automotive generator.

Answer D is incorrect. 14.6 volts is a possible peak charge voltage for an automotive generator.

24. A late model vehicle is being diagnosed for a charging problem. The generator only charges at 12.4 volts. A voltage drop test is performed on the charging output wire and 1.8 volts is measured. Technician A says that a blown fusible link in the output circuit could be the cause. Technician B says that a loose nut at the charging output connector could be the cause. Who is correct?

TASK C.6

A. A only
B. B only
C. Both A and B
D. Neither A nor B

Answer A is incorrect. A blown fusible link would have produced a higher voltage drop.

Answer B is correct. Only Technician B is correct. A loose nut at the charging output connector could cause the excessive voltage drop in the charging circuit.

Answer C is incorrect. Only Technician B is correct.

Answer D is incorrect. Technician B is correct.

25. The generator needs to be replaced on a late-model vehicle. Technician A says that the negative battery cable should be removed during this process. Technician B says the drive belt should be closely inspected during this process. Who is correct?

TASKS C.2, C.8

A. A only
B. B only
C. Both A and B
D. Neither A nor B

Answer A is incorrect. Technician B is also correct.

Answer B is incorrect. Technician A is also correct.

Answer C is correct. Both Technicians are correct. The negative battery cable needs to be disconnected while replacing the generator in order to reduce the possibility of a short to ground. The drive belt should always be closely inspected whenever replacing the generator.

Answer D is incorrect. Both Technicians are correct.

26. Technician A says that a corroded parking lamp socket could cause dimmer than normal lamp operation. Technician B says that a single filament bulb can be used in place of a dual filament bulb. Who is correct?

TASKS D.1, D.3

A. A only
B. B only
C. Both A and B
D. Neither A nor B

Answer A is correct. Only Technician A is correct. Corrosion is a form of electrical resistance and would cause the bulb in the socket to be dimmer than normal. The corrosion would act as a voltage drop and use some of the applied voltage.

Answer B is incorrect. Using a single filament bulb in place of a dual filament bulb would cause the lights to function irregularly. Any time a technician notices strange lighting operation, it is advised to check for correct bulb usage. If the corret bulb is being used and there is no problem with the bulb, then it would be advisable to check the grounds for the circuit in question.

Answer C is incorrect. Only Technician A is correct.

Answer D is incorrect. Technician A is correct.

TASK E.1

27. Which of the following electrical components would be most likely used to control the brightness of the instrument panel lights?

A. Photo resistor

B. Transducer

C. Piezo resistor

D. Rheostat

Answer A is incorrect. A photo resistor varies its resistance as the light level changes.

Answer B is incorrect. A transducer is a variable device used in pressure circuits. This device is used as a pressure sensor in many air conditioning systems.

Answer C is incorrect. Piezo resistors are used in pressure circuits such as the oil gauge. This device varies its resistance as the pressure varies.

Answer D is correct. A rheostat is a variable resistor that is used in many electrical circuits on cars and trucks. This device is the usual component used to vary the light level on the instrument panel lights.

TASK D.6

28. The under-hood light stays on continuously on a late-model vehicle. Technician A says that an open under-hood light switch is the likely cause. Technician B says that a shorted wire near the under-hood light switch could be the cause. Who is correct?

A. A only

B. B only

C. Both A and B

D. Neither A nor B

Answer A is incorrect. An open switch will not cause the under-hood light to stay on.

Answer B is correct. Only Technician B is correct. A shorted wire near the under-hood switch could cause this light to stay on continuously. The likely result of this problem would be a dead battery after the vehicle sits for a few hours.

Answer C is incorrect. Only Technician B is correct.

Answer D is incorrect. Technician B is correct.

TASKS D.4, E.4, F.9

29. Technician A says that most brake light switches are located near the brake pedal. Technician B says that many brake light switches also have circuits that activate the shift interlock system. Who is correct?

A. A only

B. B only

C. Both A and B

D. Neither A nor B

Answer A is incorrect. Technician B is also correct.

Answer B is incorrect. Technician A is also correct.

Answer C is correct. Both Technicians are correct. Brake light switches are typically located near the brake pedal. These switches also have circuits that activate the shift interlock system.

Answer D is incorrect. Both Technicians are correct.

TASKS D.3, D.4

30. The right turn signal indicator comes on when the park lights are turned on. Technician A says that a bad ground at the right front park and turn signal lamp socket could be the cause. Technician B says that a single filament bulb in a dual filament socket could be the cause. Who is correct?

   A. A only
   B. B only
   C. Both A and B
   D. Neither A nor B

   Answer A is incorrect. Technician B is also correct.

   Answer B is incorrect. Technician A is also correct.

   Answer C is correct. Both Technicians are correct. A bad ground in park and turn signal light circuits can sometimes cause an electrical back-feeding problem, which could cause the turn indicator to come on when the park lamps are turned on. Installing a single filament bulb in a dual filament socket could also cause unusual lighting problems.

   Answer D is incorrect. Both Technicians are correct.

TASKS D.4, F.9

31. Inoperative back-up lights are being diagnosed. Technician A says that a faulty multi-function switch could be the cause. Technician B says that a stuck open back-up lamp switch could be the cause. Who is correct?

   A. A only
   B. B only
   C. Both A and B
   D. Neither A nor B

   Answer A is incorrect. The back-up light circuit does not pass through the multi-function switch.

   Answer B is correct. Only Technician B is correct. A stuck open back-up lamp switch could cause the back-up lights to be inoperative. A good switch should open and close as it is operated.

   Answer C is incorrect. Only Technician B is correct.

   Answer D is incorrect. Technician B is correct.

TASK E.3

32. The oil pressure gauge intermittently moves to the area above the high setting. Technician A says that the oil pressure should be checked with a manual gauge to verify oil pressure. Technician says that an open wire in the oil gauge could be the cause of this problem. Who is correct?

   A. A only
   B. B only
   C. Both A and B
   D. Neither A nor B

   Answer A is correct. Only Technician A is correct. It would be wise to verify the oil pressure with a manual gauge whenever diagnosing an oil pressure gauge concern.

   Answer B is incorrect. An open wire would not cause an intermittent problem.

   Answer C is incorrect. Only Technician A is correct.

   Answer D is incorrect. Technician A is correct.

TASK E.3

33. A vehicle with electronic instrumentation has gauge accuracy problems. Which of the following actions would be the most likely method to diagnose the fault?

A. Swap the panel with a like-new one.

B. Replace the suspect sender unit(s).

C. Ground the sender wire at the suspect sender unit(s).

D. Connect a scan tool and compare display to gauge readings.

Answer A is incorrect. While swapping panels may help to locate a defective gauge, it is not good practice because electrical/electronic components are not returnable.

Answer B is incorrect. It is never advisable to replace components without first diagnosing the faults.

Answer C is incorrect. It is not recommended to ground sending units as a general practice. This could damage some electronic instrument clusters.

Answer D is correct. By using the scan tool, a technician can troubleshoot the problem using the OEM recommended procedure. If the diagnostic tool shows normal readings, then the gauge would be suspect.

TASKS E.3, E.5

34. The fuel gauge reads empty at all times. All of the other instrument panel gauges work correctly. Technician A says that the fuel sending unit wire could be shorted to ground. Technician B says that the gauges fuse could be blown. Who is correct?

A. A only

B. B only

C. Both A and B

D. Neither A nor B

Answer A is correct. Only Technician A is correct. A shorted sending unit wire could cause the fuel gauge to read empty all of the time. This would be true for an electromagnetic style gauge.

Answer B is incorrect. A blown gauges fuse would cause all of the gauges to be inoperative.

Answer C is incorrect. Only Technician A is correct.

Answer D is incorrect. Technician A is correct.

TASK E.5

35. The temperature sending unit needs to be replaced on a late-model vehicle. Technician A says that this repair can be made without draining the whole cooling system. Technician B says that the instrument cluster should be disconnected during this process. Who is correct?

A. A only

B. B only

C. Both A and B

D. Neither A nor B

Answer A is correct. Only Technician A is correct. The whole cooling system would not have to be drained to replace a temperature sending unit.

Answer B is incorrect. There is no need to disconnect the instrument cluster while replacing the sending unit.

Answer C is incorrect. Only Technician A is correct.

Answer D is incorrect. Technician A is correct.

36. The indicator for the bright lights does not work but the bright and dim headlights function correctly. Which of the following conditions would be the most likely cause of this problem?

TASK E.4

A. Headlight switch

B. Faulty bulb socket for the bright indicator

C. Dimmer switch

D. Multi-function switch

Answer A is incorrect. A headlight switch problem would cause all of the lights to malfunction.

Answer B is correct. A faulty bulb socket for the bright indicator could cause this problem.

Answer C is incorrect. A dimmer switch problem would cause a problem with the bright or dim headlights.

Answer D is incorrect. A multi-function switch problem would cause a problem with the bright or dim headlights.

37. The door ajar chime sounds at times at the vehicle is driven over rough roads. Technician A says that a misadjusted door striker could be the cause. Technician B says that a wire for one of the doors could be rubbing a metal bracket. Who is correct?

TASK E.4

A. A only

B. B only

C. Both A and B

D. Neither A nor B

Answer A is incorrect. Technician B is also correct.

Answer B is incorrect. Technician A is also correct.

Answer C is correct. Both Technicians are correct. A misadjusted door could cause the door ajar chime to sound when driving over rough roads by intermittently causing the door switch to open and close. A wire that is rubbing a bracket could also cause this intermittent problem.

Answer D is incorrect. Both Technicians are correct.

38. The horn blows intermittently on a late-model vehicle. Which of the following conditions would be the LEAST LIKELY cause of this problem?

TASK F.9

A. Wire rubbing a ground near the base of the steering column

B. Faulty clock spring

C. Faulty horn switch

D. Horn

Answer A is incorrect. A wire rubbing a ground near the steering column could cause the horn to blow intermittently.

Answer B is incorrect. A faulty clock spring could cause the horn to blow intermittently.

Answer C is incorrect. A faulty horn switch could cause the horn to blow intermittently.

Answer D is correct. A horn fault would not likely cause an intermittent condition of the horn sounding.

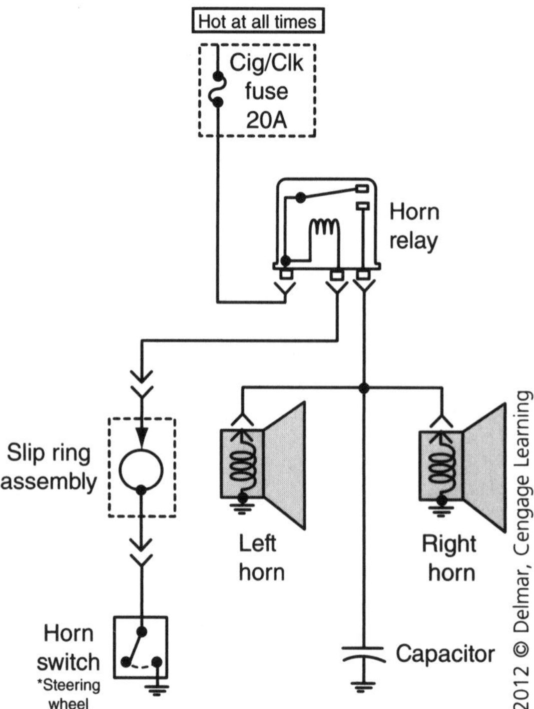

TASK E.4

39. Referring to the figure above, Technician A says that the horns are wired in series with each other. Technician B says that the horn switch provides ground to the relay coil when the switch is depressed. Who is correct?

A. A only

B. B only

C. Both A and B

D. Neither A nor B

Answer A is incorrect. The horns in the figure are wired in parallel with each other.

Answer B is correct. Only Technician B is correct. The horn switch controls the relay by providing a ground for the relay coil when the switch is activated.

Answer C is incorrect. Only Technician B is correct.

Answer D is incorrect. Technician B is correct.

40. The windshield washer motor is inoperative on a late-model vehicle. A voltage test was performed at the pump motor while the washer switch was depressed and zero volts was measured. What is the most likely cause of this problem?

TASK F.10

A. A shorted washer pump switch
B. A shorted park switch
C. An open wire in the washer pump circuit
D. A faulty washer pump motor

Answer A is incorrect. A shorted washer pump switch would cause the washer pump to receive voltage constantly.

Answer B is incorrect. A shorted wiper park switch would cause the wipers to run continuously after turning the switch off.

Answer C is correct. An open wire in the washer pump circuit could cause the washer pump to not receive any voltage when the switch is depressed.

Answer D is incorrect. The test showed zero volts being supplied to the motor so there is no evidence that the motor is faulty.

41. The power windows work from the master switch but do not work from any of the other switches in the vehicle. Technician A says that the lockout switch could be stuck closed. Technician B says that the master switch main ground connection could be disconnected. Who is correct?

TASK F.1

A. A only
B. B only
C. Both A and B
D. Neither A nor B

Answer A is incorrect. A stuck closed lockout switch would only cause the problem of the driver losing the function of turning all of the passenger switches off.

Answer B is incorrect. If the main ground was disconnected from the master switch, the whole power window system would be inoperative. This ground serves as the main ground for all of the power windows.

Answer C is incorrect. Neither Technician is correct.

Answer D is correct. Neither Technician is correct. A stuck closed lockout switch would remove the driver's ability to shut off the passenger window switches. A missing main ground at the master switch would cause all of the power window functions to be inoperative.

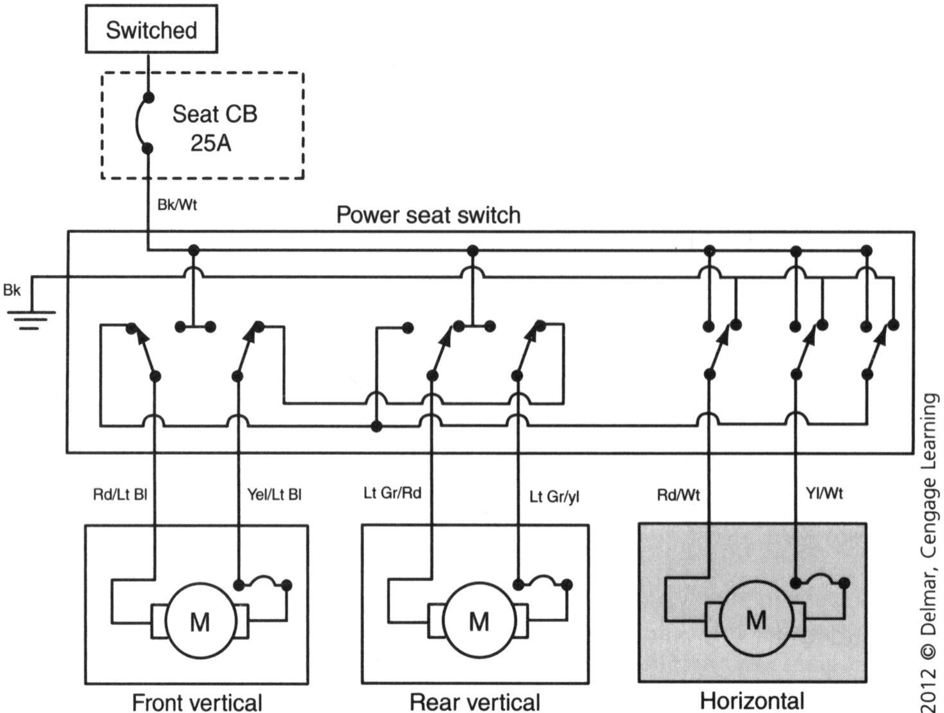

TASK F.2

42. Referring to the figure above, the horizontal seat motor is inoperative. Technician A says that the Rd/Wt wire could be broken. Technician B says that power seat switch could have a broken contact. Who is correct?

A. A only
B. B only
C. Both A and B
D. Neither A nor B

Answer A is incorrect. Technician B is also correct.

Answer B is incorrect. Technician A is also correct.

Answer C is correct. Both Technicians are correct. A broken Rd/Wt wire would cause the horizontal motor to be inoperative. In addition, a broken contact in the power seat switch could cause the horizontal seat motor to be inoperative.

Answer D is incorrect. Both Technicians are correct.

43. The rear defogger grid clears in all areas of the rear window except for one section. Technician A says that a voltmeter should be used to check for a bad ground connection. Technician B says that a voltmeter should be used to check for an open circuit in the grid. Who is correct?

TASKS A.3, F.3

A. A only

B. B only

C. Both A and B

D. Neither A nor B

Answer A is incorrect. A bad ground would have an effect on the whole rear window area, not just one section.

Answer B is correct. Only Technician B is correct. A voltmeter should be used to find a probable open circuit in the window grid. The voltage should steadily decrease as the positive lead is moved across the grid toward the ground side. On a segment with an open circuit, the voltage will stay at system level and then go to zero after the open circuit.

Answer C is incorrect. Only Technician B is correct.

Answer D is incorrect. Technician B is correct.

44. The rear defogger does not clear the glass at all when in operation. Technician A says that the rear defogger relay could be defective. Technician B says that one of the heat strips could have a break in it. Who is correct?

TASK F.3

A. A only

B. B only

C. Both A and B

D. Neither A nor B

Answer A is correct. Only Technician A is correct. A defective rear defogger relay could cause the system to be totally inoperative.

Answer B is incorrect. An open heat strip would just affect one section of the rear defogger. All of the rest of the strips should not be affected by an open in just one.

Answer C is incorrect. Only Technician A is correct.

Answer D is incorrect. Technician A is correct.

45. A radio assembly is being replaced on a late-model vehicle. Technician A says that the anti-theft unlock code may have to be programmed into the radio before it will function. Technician B says that care should be taken to prevent static electricity when handling the radio assembly. Who is correct?

TASK F.8

A. A only

B. B only

C. Both A and B

D. Neither A nor B

Answer A is incorrect. Technician B is also correct.

Answer B is incorrect. Technician A is also correct.

Answer C is correct. Both Technicians are correct. Many radios have an anti-theft function built into them and require an unlock code to be programmed into them after restoring battery power. Care should be taken to prevent static electricity from damaging the radio assembly.

Answer D is incorrect. Both Technicians are correct.

TASKS
F.7, F.8

46. Technician A says that a radio antenna can be checked with an ammeter. Technician B says that the radio antenna coaxial cable can be checked with an ohmmeter. Who is correct?

A. A only
B. B only
C. Both A and B
D. Neither A nor B

Answer A is incorrect. An ohmmeter would be used to check a radio antenna.

Answer B is correct. Only Technician B is correct. The radio antenna coaxial cable can be checked with an ohmmeter.

Answer C is incorrect. Only Technician B is correct.

Answer D is incorrect. Technician B is correct.

TASKS
A.5, F.9, F.10

47. The cruise control is inoperative on a late-model vehicle. Technician A says that a scan tool can be used to troubleshoot many of the cruise control components. Technician B says that a misadjusted brake switch could be the cause. Who is correct?

A. A only
B. B only
C. Both A and B
D. Neither A nor B

Answer A is incorrect. Technician B is also correct.

Answer B is incorrect. Technician A is also correct.

Answer C is correct. Both Technicians are correct. A scan tool can be used on many late model vehicles to troubleshoot the cruise control. Scan tools can retrieve data and trouble codes from many computerized systems on the vehicle. Scan tools can also be used to perform output function tests on electronic systems. A misadjusted brake switch could cause the cruise control system to be inoperative.

Answer D is incorrect. Both Technicians are correct.

TASKS
A.5, F.5

48. A late-model vehicle is being diagnosed with an anti-theft problem. Technician A says that many ignition keys have special circuitry that has to be replaced at regular intervals. Technician B says that a scan tool can be used to view sensor data associated with the anti-theft system. Who is correct?

A. A only
B. B only
C. Both A and B
D. Neither A nor B

Answer A is incorrect. Many late model vehicles have very intricate anti-theft systems. The keys for these systems have electronic circuitry that allows the vehicle to start. These keys do not require any regular service procedures.

Answer B is correct. Only Technician B is correct. Scan tools can be used to view sensor data associated with the anti-theft system.

Answer C is incorrect. Only Technician B is correct.

Answer D is incorrect. Technician B is correct.

49. All of the following practices are common when diagnosing supplemental inflatable restraint systems EXCEPT:

TASK F.10

A. Measuring resistance of the inflator module
B. Using a scan tool to retrieve trouble codes
C. Using a load tool to check for shorts and opens in the inflator module
D. Removing the negative battery cable prior to working around inflator devices

Answer A is correct. A technician should never attempt to use an ohmmeter on an inflator module. This practice could cause the airbag assembly to blow up due to the voltage caused by the ohmmeter.

Answer B is incorrect. It is a common practice to use a scan tool to retrieve trouble codes on an airbag system.

Answer C is incorrect. It is common to use load tool in place of the inflator modules to check for opens and short problems in the module.

Answer D is incorrect. It is common to remove the negative battery cable prior to working around the inflator devices. It is also advisable to wait about ten minutes after removing the cable so the system totally powers down.

50. Technician A says that all wire repairs on the starting system must be done with weather resistant methods. Technician B says that the wires should be voltage drop tested after repairs have been made. Who is correct?

TASKS B.8, B.9

A. A only
B. B only
C. Both A and B
D. Neither A nor B

Answer A is incorrect. Technician B is also correct.

Answer B is incorrect. Technician A is also correct.

Answer C is correct. Both Technicians are correct. Starter circuit wire repair should always be done with weather resistant repair methods to prevent corrosion from occurring. It is a good practice to perform voltage drop tests on wire repairs to test the quality of the repair.

Answer D is incorrect. Both Technicians are correct.

PREPARATION EXAM 6—ANSWER KEY

1. A
2. A
3. A
4. B
5. A
6. C
7. A
8. B
9. B
10. D
11. A
12. C
13. C
14. C
15. C
16. B
17. A
18. C
19. B
20. B
21. C
22. C
23. C
24. D
25. C
26. A
27. A
28. D
29. C
30. A
31. D
32. C
33. C
34. B
35. C
36. C
37. D
38. C
39. C
40. C
41. D
42. B
43. A
44. B
45. C
46. B
47. C
48. D
49. A
50. A

PREPARATION EXAM 6—EXPLANATIONS

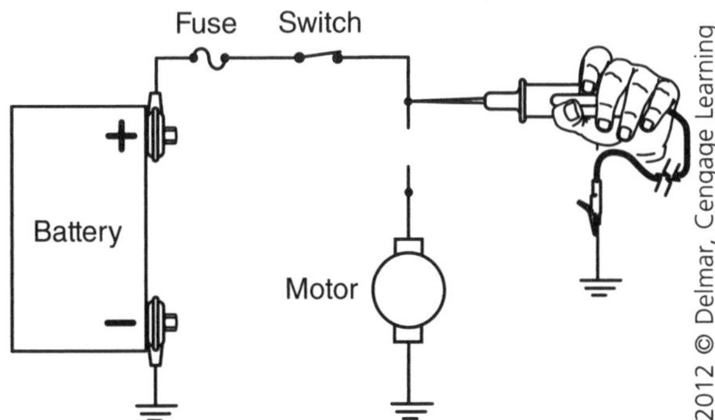

1. Referring to the figure above, Technician A says that the test light will light because it is connected before the open circuit. Technician B says that the test light would be dim because of the open circuit. Who is correct?

   A. A only
   B. B only
   C. Both A and B
   D. Neither A nor B

   Answer A is correct. Only Technician A is correct. The test light will light because voltage will be available before the open circuit.

   Answer B is incorrect. There will be source voltage available at that point and the test light will be bright.

   Answer C is incorrect. Only Technician A is correct.

   Answer D is incorrect. Technician A is correct.

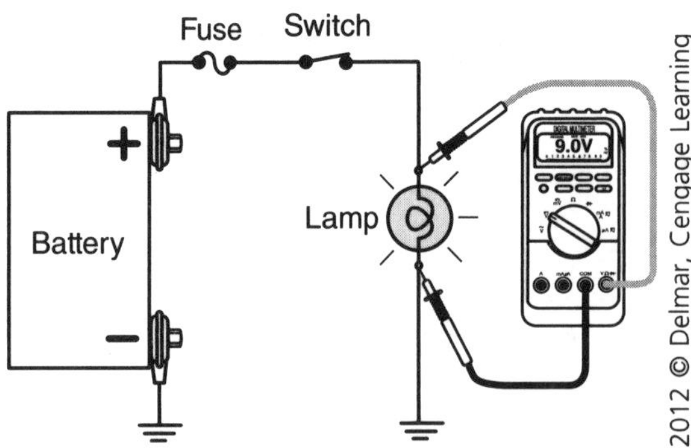

TASK A.1

2. The circuit in the figure above is a 12 volt circuit and the battery is fully charged. Technician A says that the fuse panel could have burnt contacts. Technician B says that the bulb could be open. Who is correct?

A. A only
B. B only
C. Both A and B
D. Neither A nor B

Answer A is correct. Only Technician A is correct. Burnt contacts at the fuse panel could cause low voltage to be available at the lamp. The burnt contacts would be an unwanted resistance that would drop some voltage.

Answer B is incorrect. An open bulb would not cause the low voltage drop. The circuit is a 12 volt circuit and the lamp should drop about 12 volts if everything is working as designed.

Answer C is incorrect. Only Technician A is correct.

Answer D is incorrect. Technician A is correct.

TASK A.1

3. Technician A says that a high impedance meter should be used when measuring voltages in electronic circuits. Technician B says that a digital meter with at least 1,000 megohms capacity should be used on electronic circuits. Who is correct?

A. A only
B. B only
C. Both A and B
D. Neither A nor B

Answer A is correct. Only Technician A is correct. It is safe and wise to use a high impedance meter when measuring electronic circuits.

Answer B is incorrect. Digital meters should have at least 10 megohms of impedance if electronic circuits are going to be tested.

Answer C is incorrect. Only Technician A is correct.

Answer D is incorrect. Technician A is correct.

4. Technician A says that using an amp clamp to measure current flow in an electrical circuit will interrupt the circuit. Technician B says that some low amp probes can measure milliamps with accuracy. Who is correct?

TASK A.2

A. A only
B. B only
C. Both A and B
D. Neither A nor B

Answer A is incorrect. Using an amp clamp to measure current does not interrupt the circuit. The clamp is clipped around the wire and it picks up the magnetic field in the wire and calculates the output in amps.

Answer B is correct. Only Technician B is correct. Some amp clamps have very sensitive capacity and can measure current at a very small level. These amp clamps are sometimes called low amp probes.

Answer C is incorrect. Only Technician B is correct.

Answer D is incorrect. Technician B is correct.

5. Technician A says that an electrical switch that has continuity will allow current to flow when the switch is closed. Technician B says that a piece of wire that has high resistance will have increased current flow. Who is correct?

TASK A.3

A. A only
B. B only
C. Both A and B
D. Neither A nor B

Answer A is correct. Only Technician A is correct. A closed switch should allow current to flow.

Answer B is incorrect. A wire with added resistance will have reduced current flow.

Answer C is incorrect. Only Technician A is correct.

Answer D is incorrect. Technician A is correct.

6. Technician A says that an oscilloscope can be used to view a faulty signal from a potentiometer. Technician B says that an oscilloscope can be used to view a glitch created by a permanent magnet (PM) generator. Who is correct?

TASK A.4

A. A only
B. B only
C. Both A and B
D. Neither A nor B

Answer A is incorrect. Technician B is also correct.

Answer B is incorrect. Technician A is also correct.

Answer C is correct. Both Technicians are correct. An oscilloscope displays voltage signals on a digital screen. These tools are very useful when searching for signal problems from sensors. A faulty potentiometer or PM generator could be located by using an oscilloscope.

Answer D is incorrect. Both Technicians are correct.

TASKS
A.5, F.1

7. A vehicle is in the repair shop with an inoperative power trunk release system. A scan tool is connected and the power trunk release operates when a functional test is performed. Technician A says that a faulty trunk release switch could be the problem. Technician B says that a poor connection at the trunk release motor could be the problem. Who is correct?

A. A only
B. B only
C. Both A and B
D. Neither A nor B

Answer A is correct. Only Technician A is correct. A faulty trunk release switch could cause this problem. The scan tool functional test proves that the output side of the circuit is functional.

Answer B is incorrect. The trunk release would not have worked with the scan tool output test if there was a poor connection at the trunk release motor.

Answer C is incorrect. Only Technician A is correct.

Answer D is incorrect. Technician A is correct.

TASKS
A.6, F.4

8. Which of the following tests would be LEAST LIKELY done by using a fused jumper wire?

A. Supplying power to a blower motor
B. Reference voltage supplied to a speed sensor
C. Bypassing a horn relay
D. Supplying power to a power window motor

Answer A is incorrect. A fused jumper wire could be used to supply power to a blower motor to see if it would operate.

Answer B is correct. A jumper wire should never be used to test reference voltage at a speed sensor. A high impedance voltmeter should be used when checking circuits that are associated with computers.

Answer C is incorrect. A fused jumper wire could be used to bypass a horn relay to see if the horn would blow. This test would prove that the "load side" of the horn relay was functional if the horn does blow when bypassed.

Answer D is incorrect. A fused jumper wire could be used to supply power to a power window motor. If the power window motor operates during this test, the technician knows that the motor is functional.

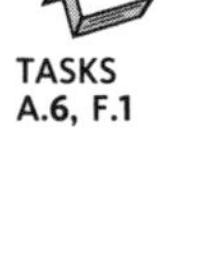

TASKS
A.6, F.1

9. A vehicle is in the repair shop with an inoperative power seat. During diagnosis, a blown power seat fuse is located. Technician A says that a short to ground between the motor and the motor ground could be the cause. Technician B says that a binding power seat motor could be the cause. Who is correct?

A. A only
B. B only
C. Both A and B
D. Neither A nor B

Answer A is incorrect. A short to ground in the ground side of the circuit would not cause increased current flow.

Answer B is correct. Only Technician B is correct. A physically binding power seat motor could cause a blown fuse. Many manufacturers use circuit breakers instead of fuses on power seats to prevent momentary surges from blowing the fuse.

Answer C is incorrect. Only Technician B is correct.

Answer D is incorrect. Technician B is correct.

10. Technician A says that a broken wire to a door jam switch will likely cause a parasitic draw problem. Technician B says that a blown horn fuse will likely cause a parasitic draw problem Who is correct?

TASK A.7

A. A only
B. B only
C. Both A and B
D. Neither A nor B

Answer A is incorrect. A broken wire to a door jam switch would not likely cause a parasitic draw problem. The dome light would not function with that door open.

Answer B is incorrect. A blown horn fuse would not likely cause a parasitic draw problem.

Answer C is incorrect. Neither Technician is correct.

Answer D is correct. Neither Technician is correct. A broken wire at a door jam switch would cause the dome light to not function when that door is opened. A blown horn fuse would cause an inoperative horn. Neither of these problems would cause a parasitic draw problem.

11. Technician A says that a fusible link will sometimes burn into two pieces when a short circuit occurs. Technician B says that a fusible link can be tested with an ammeter without disconnecting it from the vehicle. Who is correct?

TASK A.8

A. A only
B. B only
C. Both A and B
D. Neither A nor B

Answer A is correct. Only Technician A is correct. A fusible link is a circuit protection device that is used on many vehicles. It works by heating up and opening the circuit when too much current flows. Fusible links sometimes burn into two pieces when a short circuit happens.

Answer B is incorrect. A voltmeter should be used to test a fusible link while still on the vehicle.

Answer C is incorrect. Only Technician A is correct.

Answer D is incorrect. Technician A is correct.

12. Technician A says that a wiring diagram uses schematic symbols to represent electrical components. Technician B says that wiring diagrams usually show the splice and connector numbers for each circuit. Who is correct?

TASK A.9

A. A only
B. B only
C. Both A and B
D. Neither A nor B

Answer A is incorrect. Technician B is also correct.

Answer B is incorrect. Technician A is also correct.

Answer C is correct. Both Technicians are correct. Wiring diagrams use symbols to represent electrical components. In addition, wiring diagrams display the splice and connector numbers that are related to the circuit.

Answer D is incorrect. Both Technicians are correct.

TASK A.10

13. All of the following electrical tools could be used to diagnose a data bus network problem EXCEPT:

    A. Oscilloscope
    B. Digital multimeter
    C. Analog voltmeter
    D. Scan tool

Answer A is incorrect. An oscilloscope could be used to view the electrical signals being transmitted on the data bus network.

Answer B is incorrect. A digital multimeter could be used to measure the voltage levels on the data bus network.

Answer C is correct. An analog voltmeter should not be used because most analog meters do not have high impedance.

Answer D is incorrect. A scan tool could be used to communicate with the computers that use the data bus network.

TASK B.1

14. Technician A says that a 12 volt battery that has 12.2 volts at the posts is 50 percent charged. Technician B says that a 12 volt battery that has 12.6 volts at the posts is fully charged. Who is correct?

    A. A only
    B. B only
    C. Both A and B
    D. Neither A nor B

Answer A is incorrect. Technician B is also correct.

Answer B is incorrect. Technician A is also correct.

Answer C is correct. Both Technicians are correct. A fully charged battery has 2.1 volts for each of the six cells which equals 12.6 volts. The cell voltage drops as electricity is pulled from the battery. When the battery has 12.2 volts, it is approximately 50 percent charged.

Answer D is incorrect. Both Technicians are correct.

TASK B.2

15. A battery load test has been performed on an automotive battery. The voltage at the end of the 15 second test was 7.5 volts. Technician A says that the battery failed the test and should be replaced. Technician B says that extreme care should be taken when performing this test. Who is correct?

    A. A only
    B. B only
    C. Both A and B
    D. Neither A nor B

Answer A is incorrect. Technician B is also correct.

Answer B is incorrect. Technician A is also correct.

Answer C is correct. Both Technicians are correct. A battery that drops to 7.5 volts during the load test is faulty and needs to be replaced. A battery must maintain at least 9.6 volts to pass the load test. In addition, the technician should be very careful when performing this test due to the high current being drawn from the battery during the test.

Answer D is incorrect. Both Technicians are correct.

16. Technician A says that all replacement batteries should be slow charged for one hour prior to installing into the vehicle. Technician B says that the battery cable terminals should be cleaned and protected when installing a replacement battery. Who is correct?

TASKS B.4, B.5

A. A only
B. B only
C. Both A and B
D. Neither A nor B

Answer A is incorrect. A replacement battery should not have to be charged prior to installing into the vehicle, although the terminal voltage should be checked.

Answer B is correct. Only Technician B is correct. It is a good practice to clean and protect the battery cable terminals when replacing the battery.

Answer C is incorrect. Only Technician B is correct.

Answer D is incorrect. Technician B is correct.

17. Technician A says that eye protection should be worn when jumpstarting a dead battery with a booster battery. Technician B says that the last connection should be at the positive post of the dead battery when jumpstarting. Who is correct?

TASK B.6

A. A only
B. B only
C. Both A and B
D. Neither A nor B

Answer A is correct. Only Technician A is correct. It is a recommended practice to wear eye protection when performing any battery related task.

Answer B is incorrect. The last connection should be at the engine block of the dead battery vehicle when connecting booster cables to jumpstart a battery. This eliminates arcing near the battery, preventing battery explosion.

Answer C is incorrect. Only Technician A is correct.

Answer D is incorrect. Technician A is correct.

18. An engine is being diagnosed for a slow crank problem. Technician A says that testing the voltage drop on the battery cables while cranking the engine will reveal cable problems. Technician B says that worn starter brushes could cause the slow crank problem. Who is correct?

TASK B.10

A. A only
B. B only
C. Both A and B
D. Neither A nor B

Answer A is incorrect. Technician B is also correct.

Answer B is incorrect. Technician A is also correct.

Answer C is correct. Both Technicians are correct. Performing voltage drop tests on battery cables while cranking the engine is an accurate method of testing the battery cables. There should not be more than 0.5 volts on either cable during this test. Worn starter brushes could cause a slow crank problem by lowering the starter current draw due to the increased electrical resistance.

Answer D is incorrect. Both Technicians are correct.

TASKS
B.10, D.6

19. A vehicle will not crank and the technician notices that the interior lights get very dim when the ignition switch is moved to the start position. The most likely cause would be which of the following?

A. Stuck closed starter relay
B. Loose battery cable connections
C. Starter mounting bolts loose
D. Stuck open ignition switch

Answer A is incorrect. A stuck closed starter relay would cause the starter to remain engaged.

Answer B is correct. Loose battery cable connections would cause a no-crank condition. This condition would also cause the interior lights to get very dim while trying to crank the engine due to the increased electrical resistance in the loose connection.

Answer C is incorrect. Loose starter mounting bolts would cause the starter to have excess noise as well as a possible slow crank problem.

Answer D is incorrect. A stuck open ignition switch would cause a no-crank problem but the lights would not get dim while trying to crank the engine.

TASK B.9

20. Technician A says that the starter solenoid pull-in winding can be tested by touching the "start" terminal and the "bat" terminal with the ohmmeter leads. Technician B says that the starter solenoid hold-in winding can be tested by touching the "start" terminal and the solenoid case with the ohmmeter leads. Who is correct?

A. A only
B. B only
C. Both A and B
D. Neither A nor B

Answer A is incorrect. The pull-in winding can be tested by connecting the ohmmeter leads to the "start" terminal and the "motor" terminal.

Answer B is correct. Only Technician B is correct. The hold-in winding case is grounded to the solenoid case. The hold-in winding can be tested by connecting the ohmmeter leads to the "start" terminal and the solenoid case.

Answer C is incorrect. Only Technician B is correct.

Answer D is incorrect. Technician B is correct.

TASK B.10

21. Which of the following conditions would be LEAST LIKELY to cause a slow crank problem?

A. A mistimed engine
B. Worn starter brushes
C. Battery with a higher CCA rating than original specifications
D. An engine with tight main bearings

Answer A is incorrect. A mistimed engine could cause a slow crank problem due to the ignition system firing at the wrong time or if the crank/cam timing was incorrect.

Answer B is incorrect. Worn starter brushes could cause a slow crank problem due to the increase electrical resistance.

Answer C is correct. A battery with a higher CCA rating would not cause a slow crank problem. Batteries with higher CCA ratings can be used if the physical dimensions allow.

Answer D is incorrect. An engine with tight main bearings could cause a slow crank problem due to the increased physical resistance.

22. Technician A says that the replacement starter assembly should be bench-tested prior to installing on the engine. Technician B says that the starter connectors and terminals should be inspected and cleaned prior to installing a replacement starter. Who is correct?

TASK B.9

A. A only
B. B only
C. Both A and B
D. Neither A nor B

Answer A is incorrect. Technician B is also correct.

Answer B is incorrect. Technician A is also correct.

Answer C is correct. Both Technicians are correct. It is a good practice to bench-test the replacement starter prior to installing into the vehicle. In addition, the starter connectors and terminals should be inspected and cleaned.

Answer D is incorrect. Both Technicians are correct.

23. Which of the following would be the most likely condition to cause the generator to overcharge?

TASK C.1

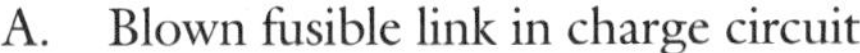

A. Blown fusible link in charge circuit
B. An open rotor coil
C. A full-fielded rotor
D. A faulty voltage regulator

Answer A is incorrect. A blown fusible link in the charge circuit would cause the generator to not charge at all.

Answer B is incorrect. An open rotor coil would cause the generator to not charge at all.

Answer C is correct. A full-fielded rotor would cause the generator to overcharge at all times.

Answer D is incorrect. A faulty voltage regulator could cause the generator to overcharge, undercharge or not charge at all.

24. A vehicle is being diagnosed for a charging problem. The generator produced 125 amps during the output test and it is rated at 130 amps. Technician A says that the generator could have a bad diode. Technician B says that the generator drive pulley could be too large in diameter. Who is correct?

TASKS C.2, C.4

A. A only
B. B only
C. Both A and B
D. Neither A nor B

Answer A is incorrect. A bad diode would cause the generator to lose more of its capacity to charge.

Answer B is incorrect. A drive pulley that is tool large in diameter would cause the generator to lose more of its capacity to charge.

Answer C is incorrect. Neither Technician is correct.

Answer D is correct. Neither Technician is correct. This generator is charging to nearly its total capacity during the output test. As long as the generator charges within approximately 10 percent of its rating, then there is nothing wrong.

TASK C.6

25. A late model vehicle is being diagnosed for a charging problem. The generator only charges at 12.4 volts. Technician A says that the charging output wire may have an excessive voltage drop. Technician B says that the charging ground circuit may have an excessive voltage drop. Who is correct?

A. A only
B. B only
C. Both A and B
D. Neither A nor B

Answer A is incorrect. Technician B is also correct.

Answer B is incorrect. Technician A is also correct.

Answer C is correct. Both Technicians are correct. Excessive voltage drop in either the positive or ground-side could cause a generator output problem. The maximum voltage drop allowable is .5 volts in either side of the circuit while charging at a high rate.

Answer D is incorrect. Both Technicians are correct.

TASK C.7

26. The wiring and connections of the charging system should be checked in all of the following ways during generator replacement EXCEPT:

A. Inspect the stator resistance.
B. Inspect the connections for tightness.
C. Inspect the wire insulation for cuts and cracks.
D. Inspect the routing of the wires and harnesses.

Answer A is correct. The stator resistance would not be checked during generator replacement.

Answer B is incorrect. The charging system connections should be checked for tightness to assure that they are secure.

Answer C is incorrect. The charging system wires should be checked for cuts and cracks to assure that they are not damaged.

Answer D is incorrect. The charging system wires should be checked for the correct routing to make sure that they do not get stretched or damaged.

TASK C.8

27. The generator needs to be replaced on a late-model vehicle. Technician A says that the replacement generator should have the same diameter pulley. Technician B says that the replacement generator should be disassembled and tested prior to installing on the vehicle. Who is correct?

A. A only
B. B only
C. Both A and B
D. Neither A nor B

Answer A is correct. Only Technician A is correct. The replacement generator should be identical to the unit being replaced. The technician should make sure that the pulley is the same diameter as well as have the same number of grooves for the belt.

Answer B is incorrect. It is not necessary to disassemble and test the replacement generator before installing on the vehicle.

Answer C is incorrect. Only Technician A is correct.

Answer D is incorrect. Technician A is correct.

28. The right side headlight of a vehicle is very dim and the left side headlight is normal. Technician A says that the dimmer switch is likely defective. Technician B says that the left side headlight could have a bad ground. Who is correct?

TASK D.1

A. A only
B. B only
C. Both A and B
D. Neither A nor B

Answer A is incorrect. If the dimmer switch is faulty, it would affect both sides.

Answer B is incorrect. A ground on the opposite side would not affect the side with the problem.

Answer C is incorrect. Neither Technician is correct.

Answer D is correct. Neither Technician is correct. A faulty dimmer would affect either the high or low beams, not just one side. A bad ground on the left side would not likely affect the headlight on the right side.

29. Technician A says that the daytime running lights utilize the headlight filaments on most late-model vehicles. Technician B says that the daytime running lights will use less current than the headlights. Who is correct?

TASKS D.1, D.3

A. A only
B. B only
C. Both A and B
D. Neither A nor B

Answer A is incorrect. Technician B is also correct.

Answer B is incorrect. Technician A is also correct.

Answer C is correct. Both Technicians are correct. The daytime running lights typically utilize the headlight filaments but usually not at full intensity, which causes less current to flow.

Answer D is incorrect. Both Technicians are correct.

30. The fog lights are dimmer than normal and a voltage drop test across the bulbs reveals 10 volts. Which of the following conditions would LEAST LIKELY cause this problem?

TASK D.1

A. Faulty fog light bulbs

B. Faulty fog light switch contacts
C. Faulty fog light relay
D. Faulty connection near the fog light switch

Answer A is correct. There is no indication of faulty fog light bulbs. The problem described shows that only 10 volts is being dropped across the bulbs, which indicates a loss of voltage somewhere in the circuit.

Answer B is incorrect. A faulty fog light switch could cause an unwanted voltage drop in the circuit and create the problem that is described.

Answer C is incorrect. A fog light relay that is faulty could cause an unwanted voltage drop in the circuit and create the problem that is described.

Answer D is incorrect. A faulty connection could cause an unwanted voltage drop and create the problem that is described.

TASKS D.4, E.4, F.9

31. Which of the following problems would most likely cause the stop lights to be inoperative?

    A. Stuck closed back-up lamp switch
    B. Incorrect turn signal flasher
    C. Blown headlight fuse
    D. Blown stop light fuse

Answer A is incorrect. A stuck closed back-up lamp switch would cause the backup lights to run continuously any time the key is turned on.

Answer B is incorrect. An incorrect turn signal flasher would cause the turn signals to malfunction. Most turn signal flashers are specific for the number and type of bulbs that are used in the system.

Answer C is incorrect. A blown headlight fuse would cause problems in the headlight operation, but not in the stop light system.

Answer D is correct. A blown stop light fuse will cause the stop lights to be inoperative.

TASKS A.5, F.9

32. A vehicle is being diagnosed with a problem with the parking assist system. Technician A says a scan tool could be used to retrieve trouble codes from the control module. Technician B says that mud on the proximity sensors could cause the system to malfunction. Who is correct?

    A. A only
    B. B only
    C. Both A and B
    D. Neither A nor B

Answer A is incorrect. Technician B is also correct.

Answer B is incorrect. Technician A is also correct.

Answer C is correct. Both Technicians are correct. The parking assist system has a module that could be scanned for trouble codes with a scan tool. A proximity sensor that is blocked by mud would not function correctly.

Answer D is incorrect. Both Technicians are correct.

TASK D.8

33. Technician A says to use dielectric grease on the trailer wiring connector to reduce the chance of corrosion. Technician B says that the turn signal flasher may flash faster when the vehicle is connected to a trailer. Who is correct?

    A. A only
    B. B only
    C. Both A and B
    D. Neither A nor B

Answer A is incorrect. Technician B is also correct.

Answer B is incorrect. Technician A is also correct.

Answer C is correct. Both Technicians are correct. It is a good practice to use some dielectric grease on the trailer wiring harness connector to resist water contact and corrosion. The turn signals will flash faster when a trailer is connected unless the flasher has been upgraded to a heavy duty flasher.

Answer D is incorrect. Both Technicians are correct.

34. The oil pressure gauge intermittently moves to the "low" setting. Technician A says that the oil and filter should be changed. Technician says that a faulty oil sending could be the cause of this problem. Who is correct?

TASK E.3

A. A only
B. B only
C. Both A and B
D. Neither A nor B

Answer A is incorrect. It would not be recommended to change the engine oil and filter without performing some other diagnosis first.

Answer B is correct. Only Technician B is correct. A faulty oil sending unit could cause an intermittent problem in the oil pressure gauge operation.

Answer C is incorrect. Only Technician B is correct.

Answer D is incorrect. Technician B is correct.

35. The temperature sending unit needs to be replaced on a late-model vehicle. Technician A says that this process can be completed without draining the whole cooling system. Technician B says that the engine should be cooled down prior to performing this repair. Who is correct?

TASK E.5

A. A only
B. B only
C. Both A and B
D. Neither A nor B

Answer A is incorrect. Technician B is also correct.

Answer B is incorrect. Technician A is also correct.

Answer C is correct. Both Technicians are correct. The temperature sending unit can be replaced without draining the whole cooling system, but it is advisable to let the engine cool down before beginning the procedure.

Answer D is incorrect. Both Technicians are correct.

36. A vehicle is being diagnosed for a problem of all of the gauges being inoperative. The vehicle is equipped with an electronic cluster that has self-diagnostic capabilities. Technician A says that this problem could be diagnosed by using the self-diagnostic function of the instrument cluster. Technician B says that a scan tool should be connected to the data link connector to view the sensor data activity at the instrument cluster. Who is correct?

TASKS A.5, E.3

A. A only
B. B only
C. Both A and B
D. Neither A nor B

Answer A is incorrect. Technician B is also correct.

Answer B is incorrect. Technician A is also correct.

Answer C is correct. Both Technicians are correct. Vehicles that use instrument clusters that have self-diagnostic capabilities are known as "smart clusters." These clusters can be manipulated to display trouble codes as well as calibration procedures by pressing a series of buttons. Scan tools can also be used to communicate with these instrument clusters.

Answer D is incorrect. Both Technicians are correct.

TASK E.5

37. All of the following types of electronic devices are used as inputs to electronic gauge assemblies EXCEPT:

A. Thermistor
B. Piezo resistor
C. Rheostat
D. Actuator

Answer A is incorrect. Thermistors are commonly used as temperature sending units for temperature gauges.

Answer B is incorrect. Piezo resistors are sometimes used as sending units for oil pressure gauges.

Answer C is incorrect. Rheostats are sometimes used as sending units for the fuel gauge.

Answer D is correct. An actuator is a control device that performs some task in response to an electrical signal.

TASKS A.6, E.1

38. The charging light stays on while driving. Technician A says that a grounded wire near the generator could be the cause. Technician B says that a faulty circuit in the instrument cluster could be the cause. Who is correct?

A. A only
B. B only
C. Both A and B
D. Neither A nor B

Answer A is incorrect. Technician B is also correct.

Answer B is incorrect. Technician A is also correct.

Answer C is correct. Both Technicians are correct. A grounded wire near the generator could cause the charge light to stay on. A faulty circuit in the instrument cluster could also cause the charge light to stay on.

Answer D is incorrect. Both Technicians are correct.

TASKS E.4, F.9

39. The chime reminder for the park assist system sounds at times as the vehicle is driven over rough roads. Which of the following conditions would be the most likely cause for this problem?

A. Open wire in the park assist circuit
B. A disconnected chime module
C. Park assist wire rubbing the vehicle body
D. Park assist sensor unplugged

Answer A is incorrect. An open wire in the park assist system would likely cause the park assist system to not work correctly.

Answer B is incorrect. A disconnected chime module would never make any sound.

Answer C is correct. A park assist wire rubbing the vehicle body could cause this intermittent chime to sound.

Answer D is incorrect. A disconnected park assist sensor would cause the park assist system to not function correctly, but it would not likely cause the system to chime periodically.

40. All of the following could cause an inoperative horn EXCEPT:

A. Open horn relay coil
B. Open in the clock spring
C. Stuck closed horn relay
D. Blown horn fuse

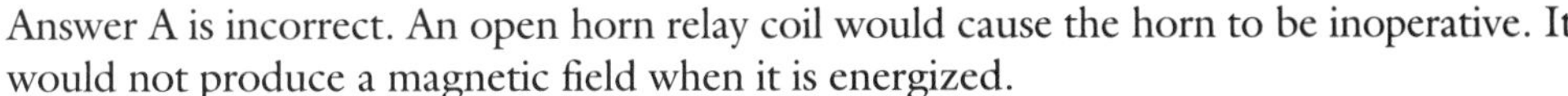

Answer A is incorrect. An open horn relay coil would cause the horn to be inoperative. It would not produce a magnetic field when it is energized.

Answer B is incorrect. An open clock spring would cause the horn to be inoperative. The clock spring is integrated into the horn control circuit on all vehicles with supplemental restraint systems (airbags).

Answer C is correct. A horn relay that sticks in the closed position would cause the horns to continuously operate.

Answer D is incorrect. A blown horn fuse would cause an inoperative horn.

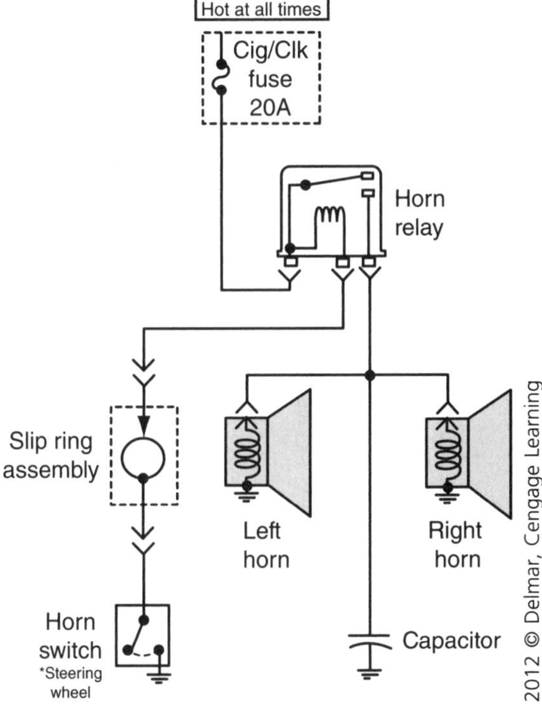

41. Referring to the figure above, all of the following conditions could cause both horns to be inoperative EXCEPT:

A. Broken wire at the relay output connection
B. Open "Cig/Clk" fuse
C. Faulty horn relay
D. Shorted horn switch

Answer A is incorrect. A broken wire at the relay output connection could cause both horns to be inoperative.

Answer B is incorrect. An open "Cig/Clk" fuse could cause both horns to be inoperative.

Answer C is incorrect. A faulty horn relay could cause both horns to be inoperative.

Answer D is correct. A shorted horn switch would cause the horn to sound all of the time.

TASK F.9

42. The horn blows intermittently while turning the steering wheel on a late-model vehicle. Which of the following conditions would be the most likely cause of this problem?
    A. Open horn circuit wire near the power distribution center
    B. Faulty clock spring
    C. Faulty horn switch
    D. Faulty horn

    Answer A is incorrect. An open horn circuit would cause the horns to be inoperative.

    Answer B is correct. A faulty clock spring could cause an intermittent ground problem that could cause the horn to blow when the steering wheel is turned.

    Answer C is incorrect. A faulty horn switch could cause the horn to blow continuously or to be inoperative. It is not likely that this problem would happen while turning the steering wheel.

    Answer D is incorrect. A faulty horn would not cause this intermittent problem that happens when the steering wheel is turned.

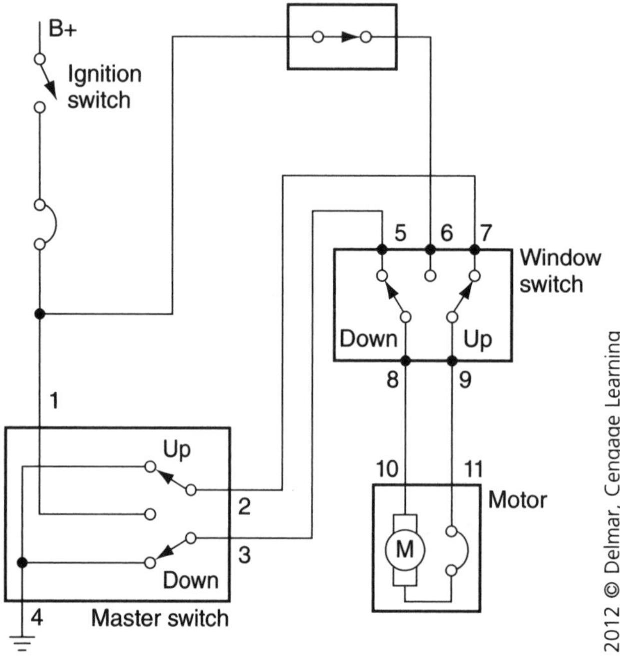

TASKS A.9, F.1

43. Referring to the figure above, the power window motor operates correctly from the window switch but does not operate at all from the master switch. Technician A says that terminal #1 could be broken. Technician B says that terminal #4 at the master switch could be broken. Who is correct?
    A. A only
    B. B only
    C. Both A and B
    D. Neither A nor B

    Answer A is correct. Only Technician A is correct. A broken terminal #1 could cause the master switch to lose its battery power source.

    Answer B is incorrect. A broken terminal #4 would cause all power window operation to be inoperative because this is the only ground connection for all of the power window circuits.

    Answer C is incorrect. Only Technician A is correct.

    Answer D is incorrect. Technician A is correct.

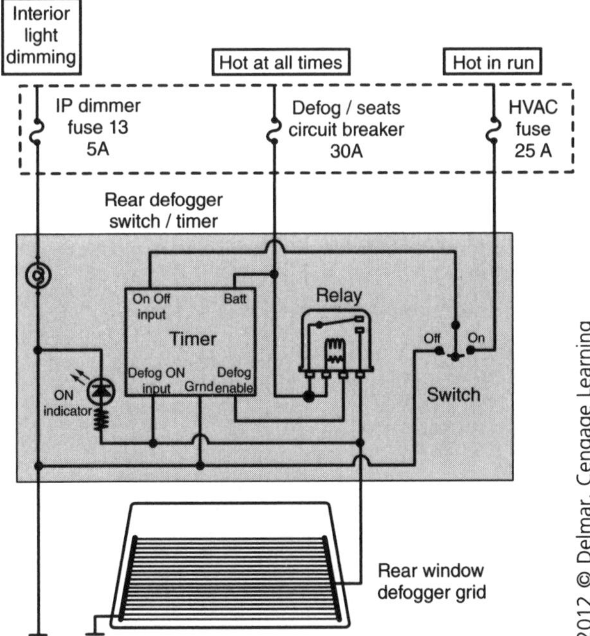

44. Referring to the figure above, the rear defogger does not function but the "on" indicator illuminates when the switch is depressed. Technician A says that the relay coil could be open. Technician B says that there could be an open wire between the relay and the rear defogger grid. Who is correct?

TASK F.3

A. A only
B. B only
C. Both A and B
D. Neither A nor B

Answer A is incorrect. The "on" indicator would not function if the relay coil was open.

Answer B is correct. Only Technician B is correct. It is possible to have an open wire between the relay output (after the splice) and the rear defogger grid since the "on" indicator comes on when the switch is activated.

Answer C is incorrect. Only Technician B is correct.

Answer D is incorrect. Technician B is correct.

45. A minivan with power sliding doors is being diagnosed for a power sliding door problem. The passenger side sliding door does not open with the handle switch, but it works correctly from the driver's switch. Technician A says that the passenger handle switch could be faulty. Technician B says that the passenger handle switch could have a bad connection. Who is correct?

TASKS F.1, F.2

A. A only
B. B only
C. Both A and B
D. Neither A nor B

Answer A is incorrect. Technician B is also correct.

Answer B is incorrect. Technician A is also correct.

Answer C is correct. Both Technicians are correct. A bad passenger handle switch or a bad connection at the passenger handle switch could cause the problem in this question. Most of the circuit is functional since the power door works correctly from the driver's switch.

Answer D is incorrect. Both Technicians are correct.

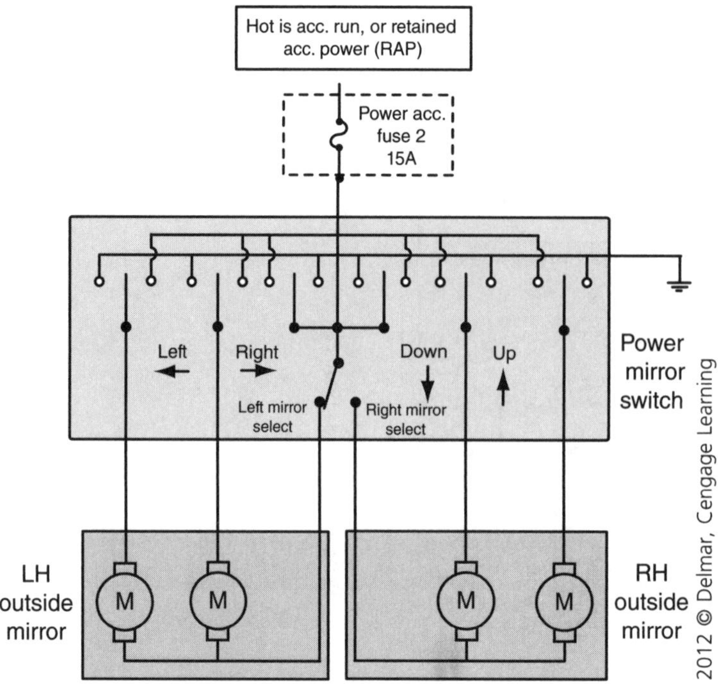

TASK F.1

46. Referring to the figure above, the right power mirror functions normally but the left side power mirror does not function in the up and down directions. Technician A says that mirror select switch could be defective. Technician B says that the built-in circuit breaker in the up/down motor could be defective. Who is correct?

A. A only
B. B only
C. Both A and B
D. Neither A nor B

Answer A is incorrect. A defective mirror select switch would prevent both of the mirrors on the left side from working.

Answer B is correct. Only Technician B is correct. A defective circuit breaker in the up/down motor could cause that motor to be inoperative.

Answer C is incorrect. Only Technician B is correct.

Answer D is incorrect. Technician B is correct.

TASK F.8

47. Technician A says that the power antenna uses a bi-directional permanent magnet motor. Technician B says that some power antenna systems use relays to control the up and down circuit operation. Who is correct?

A. A only
B. B only
C. Both A and B
D. Neither A nor B

Answer A is incorrect. Technician B is also correct.

Answer B is incorrect. Technician A is also correct.

Answer C is correct. Both Technicians are correct. Power antennas are typically controlled relays that supply power to go up and down. The motors for power antennas are typically bi-directional permanent magnet design.

Answer D is incorrect. Both Technicians are correct.

48. The auxiliary power outlet is inoperative and the fuse is found to be open. What is the LEAST LIKELY cause for this condition?

TASKS A.6, F.7

A. Shorted power wire near the auxiliary connector
B. Foreign metal object in the auxiliary power outlet
C. A faulty electrical device connected to the outlet
D. Open internal connection at the power outlet

Answer A is incorrect. A shorted power wire could cause the fuse to blow for the auxiliary power outlet.

Answer B is incorrect. A foreign metal object in the auxiliary power outlet could cause the fuse to blow.

Answer C is incorrect. A faulty electrical device that gets connected to the auxiliary power outlet could cause the fuse to blow.

Answer D is correct. An open circuit will not cause a fuse to blow.

49. Which of the following conditions would be LEAST LIKELY to cause the airbag light to stay on while the vehicle is in operation?

TASKS E.4, F.10

A. Bad connection at the airbag bulb socket
B. A shorted airbag inflator
C. A faulty seatbelt buckle
D. An open clock spring

Answer A is correct. A bad connection at the airbag bulb connection would not cause the airbag light to stay on while driving the vehicle.

Answer B is incorrect. A shorted airbag inflator would often cause the airbag light to stay on while driving the vehicle.

Answer C is incorrect. A faulty seatbelt buckle could cause the airbag light to stay on while driving the vehicle.

Answer D is incorrect. An open clock spring could cause the airbag light to stay on while driving the vehicle.

50. An airbag system needs to be disarmed during a driver's inflator module replacement. Technician A says that it is necessary to remove the negative battery cable prior to component removal. Technician B says that the inflatable devices should be laid "face down" in a secure area after being removed from a vehicle. Who is correct?

TASK F.10

A. A only
B. B only
C. Both A and B
D. Neither A nor B

Answer A is correct. Only Technician A is correct. Removing the negative battery cable is recommended prior to removing the components of the airbag system.

Answer B is incorrect. The inflatable device should be laid "face up" in a secure area after being removed from the vehicle.

Answer C is incorrect. Only Technician A is correct.

Answer D is incorrect. Technician A is correct.

SECTION

7 Appendices

PREPARATION EXAM ANSWER SHEET FORMS

ANSWER SHEET

1. ______________
2. ______________
3. ______________
4. ______________
5. ______________
6. ______________
7. ______________
8. ______________
9. ______________
10. ______________
11. ______________
12. ______________
13. ______________
14. ______________
15. ______________
16. ______________
17. ______________
18. ______________
19. ______________
20. ______________
21. ______________
22. ______________
23. ______________
24. ______________
25. ______________
26. ______________
27. ______________
28. ______________
29. ______________
30. ______________
31. ______________
32. ______________
33. ______________
34. ______________
35. ______________
36. ______________
37. ______________
38. ______________
39. ______________
40. ______________
41. ______________
42. ______________
43. ______________
44. ______________
45. ______________
46. ______________
47. ______________
48. ______________
49. ______________
50. ______________

ANSWER SHEET

1. ______________
2. ______________
3. ______________
4. ______________
5. ______________
6. ______________
7. ______________
8. ______________
9. ______________
10. ______________
11. ______________
12. ______________
13. ______________
14. ______________
15. ______________
16. ______________
17. ______________
18. ______________
19. ______________
20. ______________
21. ______________
22. ______________
23. ______________
24. ______________
25. ______________
26. ______________
27. ______________
28. ______________
29. ______________
30. ______________
31. ______________
32. ______________
33. ______________
34. ______________
35. ______________
36. ______________
37. ______________
38. ______________
39. ______________
40. ______________
41. ______________
42. ______________
43. ______________
44. ______________
45. ______________
46. ______________
47. ______________
48. ______________
49. ______________
50. ______________

ANSWER SHEET

1. ____________
2. ____________
3. ____________
4. ____________
5. ____________
6. ____________
7. ____________
8. ____________
9. ____________
10. ____________
11. ____________
12. ____________
13. ____________
14. ____________
15. ____________
16. ____________
17. ____________
18. ____________
19. ____________
20. ____________
21. ____________
22. ____________
23. ____________
24. ____________
25. ____________
26. ____________
27. ____________
28. ____________
29. ____________
30. ____________
31. ____________
32. ____________
33. ____________
34. ____________
35. ____________
36. ____________
37. ____________
38. ____________
39. ____________
40. ____________
41. ____________
42. ____________
43. ____________
44. ____________
45. ____________
46. ____________
47. ____________
48. ____________
49. ____________
50. ____________

ANSWER SHEET

1. ______________
2. ______________
3. ______________
4. ______________
5. ______________
6. ______________
7. ______________
8. ______________
9. ______________
10. ______________
11. ______________
12. ______________
13. ______________
14. ______________
15. ______________
16. ______________
17. ______________
18. ______________
19. ______________
20. ______________
21. ______________
22. ______________
23. ______________
24. ______________
25. ______________
26. ______________
27. ______________
28. ______________
29. ______________
30. ______________
31. ______________
32. ______________
33. ______________
34. ______________
35. ______________
36. ______________
37. ______________
38. ______________
39. ______________
40. ______________
41. ______________
42. ______________
43. ______________
44. ______________
45. ______________
46. ______________
47. ______________
48. ______________
49. ______________
50. ______________

ANSWER SHEET

1.	________	21.	________	41.	________
2.	________	22.	________	42.	________
3.	________	23.	________	43.	________
4.	________	24.	________	44.	________
5.	________	25.	________	45.	________
6.	________	26.	________	46.	________
7.	________	27.	________	47.	________
8.	________	28.	________	48.	________
9.	________	29.	________	49.	________
10.	________	30.	________	50.	________
11.	________	31.	________		
12.	________	32.	________		
13.	________	33.	________		
14.	________	34.	________		
15.	________	35.	________		
16.	________	36.	________		
17.	________	37.	________		
18.	________	38.	________		
19.	________	39.	________		
20.	________	40.	________		

ANSWER SHEET

1. ________	21. ________	41. ________
2. ________	22. ________	42. ________
3. ________	23. ________	43. ________
4. ________	24. ________	44. ________
5. ________	25. ________	45. ________
6. ________	26. ________	46. ________
7. ________	27. ________	47. ________
8. ________	28. ________	48. ________
9. ________	29. ________	49. ________
10. ________	30. ________	50. ________
11. ________	31. ________	
12. ________	32. ________	
13. ________	33. ________	
14. ________	34. ________	
15. ________	35. ________	
16. ________	36. ________	
17. ________	37. ________	
18. ________	38. ________	
19. ________	39. ________	
20. ________	40. ________	

Glossary

Airbag A passive restraint system having an inflatable bag located in the center of the steering wheel in front of the driver and, in later model vehicles, a second inflatable bag located in the dash in front of the front seat passenger that inflates on vehicle impact.

Alternator A belt-driven generator that converts mechanical energy to electrical energy.

Amperage The flow of electrical energy.

Antenna A wire or other conductive device used for radiating or receiving electromagnetic signals, such as those for radio, television, or radar.

Anti-Theft System A deterrent system designed to scare off would-be vehicle thieves.

Armature A part moved through a magnetic field to produce an electric current.

Back-Up Light Lamps that provide rear illumination when the vehicle is being backed up.

Battery A device that converts chemical energy into electrical energy.

Bearing A component that reduces friction between a stationary and rotating part, such as a shaft.

Body Control Module (BCM) A component of the computerized self diagnosis system.

Brake Lights Red lamps at the rear of the vehicle that are illuminated when the brake pedal is applied.

Brake Light Switch A component of the braking system that completes an electrical circuit to illuminate the brake lights when the brake pedal is applied.

Brush A conductive component that rides on the commutator or slip ring to provide an electrical circuit between rotating and stationary components.

Bulb A glass envelope containing a filament to provide illumination.

Bulkhead Connector A connector for wires that are to pass through the bulkhead of a vehicle.

Buzzer An electric sound generator that makes a buzzing noise.

Capacitor An electrical device for the temporary storage of electricity, often used to reduce RFI.

Charge The passing of an electric current through a battery to restore its energy.

Clock Spring A device used on vehicles with airbags that allows a hard-wired circuit to be connected the a rotating steering wheel.

Computer A component that receives inputs, performs calculations, and then sends outputs to load devices. Other names for this component include control module, controller, logic device, and electronic control unit.

Concealed Headlights A headlamp system that retracts the lamps into the bodywork when they are turned off.

Continuity A complete path for electrical flow.

Corrosion A chemical action that eats away material such as metal, paint, or wire.

Courtesy Lights Lamps that illuminate the interior of a vehicle when a door is opened.

Cruise Control A device that automatically maintains vehicle speed over a wide range of terrain conditions.

Current The flow of electricity, measured in amperes.

Current Draw Test A test to determine amperes required by the starter motor during starting operation.

Defogger A part of a heater system to prevent windshield or rear window fogging or icing.

Digital An electrical signal having two states, on and off.

Electric Fuel Pump An electrical device used to draw fuel from the fuel tank and deliver it to the engine.

Electronic Ignition System An ignition system controlled by solid state electrical signals.

Field Coil A coil of insulated wire, usually wound around an iron core, through which current is passed to produce a magnetic field.

Flywheel A heavy metal wheel that is attached to the crankshaft and rotates with it.

Fusible Link A bar or wire that is designed to melt due to heat when a specified current passing through it is exceeded.

Ground The path, generally the body of the vehicle, for the return of an electrical circuit. Also, a term used for causing an accidental or intentional short circuit.

Harness A group of electrical conductors.

Hazard Warning System Vehicle perimeter lighting and associated switches and wiring that flash to give warning to a hazard.

Headlamps The lamps at the front of a vehicle to provide illumination for the road ahead.

High Intensity Discharge (HID) A type of headlight that uses high voltage levels to create a very intense light.

Horn Relay An electromagnetic device used to activate the horn when the horn switch is closed.

Hydrometer An instrument used to measure the specific gravity of a liquid.

Ignition Switch The main power switch, generally key operated, of a vehicle.

Inertia Switch A switch found in the fuel pump circuit to turn off the fuel pump, and other vehicle accessories, in the event of a collision.

Interior Lights Lighting in the interior of a vehicle, often called courtesy lights.

Intermittent An event that only happens part of the time.

Light Emitting Diode (LED) An electronic device that creates light when a voltage is applied to the anode and ground is connected to the cathode. LEDs are widely used on vehicles instead of incandescent bulbs because they draw less current and last longer.

Linkage Levers or rods used to transmit power for one part to another.

Load A device connected to an electrical circuit to provide resistance and/or control the current flow.

Load Test An electrical test for motors and batteries in which current draw and voltage is measured.

Logic Device A component that receives inputs, performs calculations, and then sends outputs to load devices. Also referred to as a computer.

Module a control assembly designed to perform one or more specific tasks.

Ohm A unit of measure of electrical resistance.

On-Board Computer The resident or main computer in a vehicle.

Open Circuit An incomplete electrical circuit.

Power Steering Pump A hydraulic pump used to provide a fluid boost for ease in vehicle steering.

Printed Circuit Electrically conductive circuit paths generally etched on rigid or flexible strata.

Printed Circuit Board An insulated board on which a printed circuit is etched.

Resistance The opposition to electrical flow.

Scan Tool A computer that can be connected to a vehicle that will communicate with the on-board computers (logic devices). Scan tools can typically display live data, trouble codes, as well as perform output tests.

Schematic A drawing of a system using symbols to represent components.

Sending Unit An electrical or mechanical sensing device to transmit certain conditions to a remote meter or gauge.

Series A part of an electrical circuit whereby one component is connected to another, as negative to positive, and so on.

Servo A device that converts hydraulic pressure to mechanical movement.

Short Circuit The intentional or unintentional grounding of an electrical circuit.

Shunt A parallel electrical connection or circuit.

Solenoid A device that uses electromagnetism to create linear movement.

Speed Sensor An electrical device that senses the speed of a rotating shaft or vibrating member.

Squeal A continuous high-pitched noise.

Starter The electric motor and drive used to start a vehicle engine.

Starter Solenoid A magnetic switch used to engage the starter for starting an engine.

Switch A mechanical device used to open and close an electrical circuit.

Transmission A device used to couple a motor to a mechanical mechanism.

Troubleshoot To determine the problem, the cause of the problem, and the solution by systematic reasoning.

Turn Signal Lights on the four corners of a vehicle to signal a turn.

Voltage Electrical pressure. Also known as electromotive force.

Wiper A mechanical arm that moves back and forth over the windshield to remove water.

Notes